BETWEEN THE EARTH AND THE HEAVENS

Historical Studies in the Physical Sciences

History of Modern Physical Sciences

ISSN 1793-0820

Aims and Scope

The series will include a variety of books dealing with the development of physics, astronomy, chemistry and geology during the past two centuries (1800–2000). During this period there were many important discoveries and new theories in the physical sciences which radically changed our understanding of the natural world, at the same time stimulating technological advances and providing a model for the growth of scientific understanding in the biological and behavioral sciences.

While there is no shortage of popular or journalistic writing on these subjects, there is a need for more accurate and comprehensive treatments by professional historians of science who are qualified to discuss the substance of scientific research. The books in the series will include new historical monographs, editions and translations of original sources, and reprints of older (but still valuable) histories. Efforts to understand the worldwide growth and impact of physical science, not restricted to the traditional focus on Europe and the United States, will be encouraged. The books should be authoritative and readable, useful to scientists, graduate students and anyone else with a serious interest in the history, philosophy and social studies of science.

Published

Vol. 5 *Between the Earth and the Heavens: Historical Studies in the Physical Sciences*
by Helge Kragh (Niels Bohr Institute, Denmark)

Vol. 4 *Rocks, Radio and Radar: The Extraordinary Scientific, Social and Military Life of Elizabeth Alexander*
by Mary Harris (University College London Institute of Education, UK)

Vol. 3 *Matter and Spirit in the Universe: Scientific and Religious Preludes to Modern Cosmology*
by Helge Kragh (Aarhus University, Denmark)

Vol. 2 *Stalin's Great Science: The Times and Adventures of Soviet Physicists*
by Alexei B Kojevnikov

Vol. 1 *The Kinetic Theory of Gases: An Anthology of Classic Papers with Historical Commentary*
by Stephen G Brush (University of Maryland, USA)
edited by Nancy S Hall (University of Maryland, USA)

More information on this series can also be found at http://www.worldscientific.com/series/hmps

HISTORY OF MODERN PHYSICAL SCIENCES – VOL. 5

BETWEEN THE EARTH AND THE HEAVENS

Historical Studies in the Physical Sciences

Helge Kragh

Niels Bohr Institute, Denmark

World Scientific

NEW JERSEY • LONDON • SINGAPORE • BEIJING • SHANGHAI • HONG KONG • TAIPEI • CHENNAI • TOKYO

Published by

World Scientific Publishing Europe Ltd.

57 Shelton Street, Covent Garden, London WC2H 9HE

Head office: 5 Toh Tuck Link, Singapore 596224

USA office: 27 Warren Street, Suite 401-402, Hackensack, NJ 07601

Library of Congress Cataloging-in-Publication Data
Names: Kragh, Helge, 1944– author.
Title: Between the earth and the heavens : historical studies in the physical sciences /
Helge Kragh, Niels Bohr Institute, Denmark.
Description: New Jersey : World Scientific, [2021] | Series: History of modern physical sciences,
1793-0820 ; vol. 5 | Includes bibliographical references and index.
Identifiers: LCCN 2020057458 | ISBN 9781786349842 (hardcover) |
ISBN 9781786349859 (ebook for institutions) | ISBN 9781786349866 (ebook for individuals)
Subjects: LCSH: Physical sciences--History.
Classification: LCC Q125 .K685 2021 | DDC 500.209--dc23
LC record available at https://lccn.loc.gov/2020057458

British Library Cataloguing-in-Publication Data
A catalogue record for this book is available from the British Library.

For any available supplementary material, please visit
https://www.worldscientific.com/worldscibooks/10.1142/Q0290#t=suppl

Desk Editors: Jayanthi Muthuswamy/Michael Beale/Shi Ying Koe

Typeset by Stallion Press
Email: enquiries@stallionpress.com

Printed in Singapore

To Philip and Lucas

Preface

Few readers, whether they have a scientific background or not, will be aware of the impact of the second law of thermodynamics on the cultural, religious, and literary science during the Victorian era. Nor will they know that Darwin's theory of evolution once inspired chemical thinking, or that the long discarded word *ether* had more than a little in common with the dark energy discovered by astronomers at the end of the twentieth century. And who has ever heard of the thought experiment known as Hilbert's infinite hotel? These and many other stories are told in the present book, a collection of new essays on episodes and themes in the modern history of the physical sciences.

As far as the term "modern" is concerned, by and large, it refers to the period from the mid-nineteenth century to the early twenty-first century. From the historian's perspective, if not perhaps from the scientist's, this long period qualifies as the era of modern science. As for the term "physical sciences," it is used in more or less its standard meaning of physics and its allied sciences, which are here taken to be chemistry, astronomy, and cosmology. However, it is principally the connection between the different physical sciences which is the focus of the book that to a large extent deals with interdisciplinary aspects and how they have evolved through history. As indicated by the title, an allusion to Hamlet's famous statement to Horatio, several of the chapters refer to the historical relationship between the laboratory sciences chemistry and physics and the celestial sciences astronomy and cosmology. Physical chemistry and astrophysics are well-known cases of interdisciplinary sciences, whereas cases such as chemical physics, astrochemistry, and so-called cosmical physics are presumably less known. Mathematics is not a natural science, but it features importantly as the language of theoretical physics, which eventually also became the language of large parts of astronomy and chemistry. Philosophy is another, and very different, language or perspective, which mostly appears implicitly but is nonetheless of great importance, as becomes clear in later sections.

Contrary to what is sometimes called "scientists' history" — the triumphalist description of the past as leading smoothly and progressively toward the present — from a historical perspective fruitless speculations, controversies, and plain errors are no less interesting than the successes. They are part and parcel of science, as testified by several of the cases dealt with in the various chapters. This book is structured in four parts, essentially dealing with developments in physics, chemistry, astronomy, and cosmology, respectively. The chapters and sections included in each of the parts are cases or themes I have previously investigated in scholarly detail in academic and therefore less accessible publications. Although I have attempted to order the sections roughly chronologically, there is little unity among them and they can, in most cases, be read separately. They are not entirely separate, though, and throughout the book I have made use of cross-references in order not to miss the connections to other sections. The book is provided with a large number of quotations, the references to which are collected in an appendix except if the source of the quotation is obvious from the text. On the other hand, to ease readability I have avoided references to the comprehensive scientific literature and also to the secondary literature made up of learned books and articles.

About the Author

Helge Kragh graduated in physics and chemistry in 1970 from the University of Copenhagen and he is presently Emeritus Professor at the Niels Bohr Institute. He has formerly occupied positions as Professor of history of science and technology at Cornell University, USA, the University of Oslo, Norway, and Aarhus University, Denmark. Kragh is a member of the Royal Danish Academy of Sciences and Letters and other learned academies. He has served as Chief Editor of *Centaurus*, an international history of science journal, and was a co-founder of ESHS, the European Society for History of Science. Among his recent scientific prizes are the Physics Estoire Prize (2016), the Abraham Pais Prize for History of Physics (2019), and the Roy G. Neville Prize in Chemical Biography (2019).

Kragh's academic work focuses on the histories of physics, chemistry, astronomy, and cosmology, and also the historical relationship between science and religion. He is the author of more than 500 papers and 30 books, many of them of a scholarly nature but others more popular. Recent English-language books include *Higher Speculations: Grand Theories and Failed Revolutions in Physics and Cosmology* (Oxford University Press, 2011), *Masters of the Universe: Conversations with Cosmologists of the Past* (Oxford University Press, 2014), *Varying Gravity: Dirac's Legacy in Cosmology and Geophysics* (Springer, 2016), *From Transuranic to Superheavy Elements: A Story of Dispute and Creation* (Springer, 2018), *Ludvig Lorenz: A Nineteenth-Century Theoretical Physicist* (Royal Danish Academy of Sciences and Letters, 2018), and he also co-edited, with M. Longair, *The Oxford Handbook in the History of Modern Cosmology* (Oxford University Press, 2019).

Contents

Part 3: Astronomy and Astrophysics

Part 1

Matters of Physics

1.1

The Many Faces of Thermodynamics

1.1

The Many Faces of Thermodynamics

According to Einstein, physics develops cumulatively and evolutionarily rather than by way of drastic revolutions. He did not believe that theories of physics could be final, and that included his own masterpiece, the general theory of relativity. And yet Einstein thought that the classical theory of thermodynamics, because of its generality and logical simplicity, might be an exception. As he wrote in his autobiographical notes of 1946, "It is the only physical theory of a universal content which I am convinced that within the framework of the applicability of its basic concepts, it will never be overthrown" (Schilpp, 1949, p. 33). Einstein's high appreciation of thermodynamics, a theory completed in the late nineteenth century, is shared by almost all scientists.

From a formal point of view, thermodynamics consists of two fundamental principles, usually referred to as the first and the second law. To these are sometimes added a third law, which can be regarded as a refinement of the second law. As indicated by its name, thermodynamics has its historical roots in the recognition that heat is a form of molecular motion, which in the 1840s led to the modern concept of energy as a universally conserved quantity, thereby adding to mass conservation a new conservation law. Less than ten years later, this first law was supplemented with the equally fundamental second law, and the new concept of entropy took its place alongside the concept of energy.

The two classical laws of thermodynamics were discussed in a cosmological context that inevitably involved philosophical and religious aspects. Thermodynamics was of great interest to physicists, chemists, and engineers, and it mattered to philosophers and theologians because of its cosmological implications. Energy and entropy entered the cultural struggle in the late nineteenth century between, on the one hand, the traditional view of science, religion, and society and, on the other, the rival view held by materialists, positivists, and atheists. Did the authoritative science of thermodynamics lead to a universal "heat death" in the far future? Did it justify the belief in a world of finite age and thus a creation of the universe?

1.1.1. ENERGY CONSERVATION

Given the massive role that energy plays in all areas of modern life — economically, politically, environmentally, and so on — it may be hard to believe that the concept of energy, in our sense of the term, is less than two centuries old. During the Napoleonic era, energy was still unknown. The concept had to be invented, and the invention was intimately related to the insight dating from the 1840s that there is something in nature which is always conserved, which cannot be created and can never perish. The very name "energy" only came into general use in the 1860s, although it can be found earlier. The English natural philosopher and polymath Thomas Young used it in a lecture of 1807, but only in its more restricted meaning corresponding to a body's kinetic energy or what at the time was known as *vis viva*, either mv^2 or $\frac{1}{2}mv^2$. The word *energeia* was originally coined by Aristotle with connotations quite different from the modern one, such as "activity," "actual presence," and "being at work." Fifty years after Young's lecture, energy was on everybody's lips, except that most people called it "force" rather than energy. In this section, I use the two terms synonymously.

The discovery of the law or principle of energy conservation is a classic case in the history of nineteenth-century physical sciences.

The roads that led to the discovery in the early 1840s are complex and confusing, not least because no single scientist can be credited with the discovery. More than half a dozen scientists were involved in the discovery process, most of them working independently and following different routes. Given the fundamental nature of the law, it is noteworthy that none of the discoverers or co-discoverers were professional physicists but instead medical doctors, engineers, or amateur scientists. It is generally accepted that the principal discoverers of energy conservation were the German physician Julius Robert Mayer and the British brewer and amateur scientist James Prescott Joule. Other contributors included the Danish engineer Ludvig August Colding and the German pharmacist and chemist Karl Friedrich Mohr.

The law of energy conservation was only formulated mathematically and discussed as a unifying principle valid for all of nature in an extensive memoir of 1847, written by 26-year-old Hermann Helmholtz, at the time an army medical doctor and self-taught physicist. What is today considered one of the great classics of science had been rejected by the editor of the influential journal *Annalen der Physik und Chemie*, and for this reason Helmholtz was forced to self-publish it as a monograph. At the time, passionately in love with Olga von Velten (whom he married two years later), he originally titled his treatise *Über die Erhaltung der Kraft, eine Physikalische Abhandlung zur Belehrung seiner Theuren Olga* (On the Conservation of Force, a Physics Essay for the Instruction of His Beloved Olga). Unfortunately, the subtitle with the beautiful dedication to Olga was lost in the printed version. Apart from this curiosity, for several years Helmholtz's masterpiece was ignored by physicists as well as physiologists.

According to Mayer, who offered a first account of his idea of force or energy conservation in 1842, there exists in nature a quantity that is always conserved although it can be converted from one form to another. In a treatise of 1845, he formulated his doctrine, partly philosophical and partly scientific, as follows: "Force as a cause of motion is an indestructible entity. No effect arises without a cause. No cause disappears without a corresponding effect. … It can be proved *a priori* and confirmed everywhere by experience that the various forms of forces can be transformed into one another" (Lindsay, 1973, p. 78). Based on rather inaccurate experiments, Mayer calculated the conversion factor between mechanical work and heat energy to 3.65 calories/J. He was also the first to apply the law of energy conservation in an astronomical context, which he did by proposing a new theory for the origin of the Sun's heat (see Section 3.2.1). Joule's independent and slightly later formulation of the energy law was less philosophical than Mayer's and more solidly based on experiments. His focus was not so much on energy conservation as a general principle but on the interconversions of different forms of energy, such as electrical, mechanical, chemical, and thermal. For the mechanical equivalence of heat, his best result was 4.236 calories/J, not far from the modern value of 4.184 calories/J.

Apart from being a fundamental concept of physics, today, energy is predominantly an entity of technological, economic, environmental, and commercial interest. In the mid-nineteenth century, energy was seen in a different and broader perspective, including cultural and spiritual dimensions, which reflected the beginning controversy between traditional Christian values and the new disturbing trends of atheism, materialism, and socialism. Most of the historical actors responsible for the law of energy conservation were sincere Christians who found it natural to interpret the law apologetically, as an argument for an omnipresent and omnipotent God. This is what Joule stated on several occasions, and Mayer and several others of the early expositors of the law spoke the same language. The logic of the argument was that since energy can neither be created nor destroyed naturally, ultimately its origin must be supernatural and a sign of a divine creator maintaining the order of the universe.

Mayer was convinced that the existence and immortality of the soul followed from energy conservation and that the new science of thermodynamics provided an effective weapon in the fight against godless materialism. The same viewpoint was expressed by Colding, whose idea of imperishable forces was directly inspired by his fundamentalist Christian faith. Colding tended to believe that the forces of nature were immaterial and included intellectual or mental activities. Only by including the mental forces would the law of energy conservation be strictly valid. In an address of 1856 to the Royal Danish Academy of Sciences and Letters, Colding said:

> "The growth of every new force necessarily demands other forces for its nourishment and support. ... I do not believe I am mistaken in expressing the view that it is the forces of nature, in their many forms, which serve as support for the intellect, and that intellectual activity evolves at the expense of these. ... As the intellectual life evolves at a rapid pace, the abundance of forces of nature must be in a continuous decline, because the sum of all these forces is the invariable quantity originally created by God!" (Dahl, 1972, p. 126).

Colding's version of the energy law was unorthodox, and although a few other scientists shared his view about mental forces, it was generally agreed that only measurable and lifeless forces were real and belonged to the so-called natural forces.

It is one of the many ironies of history that what was at first considered scientific support for religious belief was subsequently appropriated by the opposite camp and presented as an argument for materialism and atheism. The premise of the latter argument, apart from the belief in absolute energy conservation, was that the world can only be explained naturalistically, meaning that transcendental and supernatural causes are excluded. From this premise, materialist thinkers concluded that the world has existed in an infinity of time and will last for another infinity. There is no need for an initial creation of matter and energy, and therefore also no need for a divine creator.

This was precisely the message of the German medical doctor and science popularizer Ludwig Büchner, a leading materialist and freethinker. His widely read and frequently reprinted book of 1855 with the title *Kraft und Stoff* (Force and Matter) has been called the gospel of materialism. Büchner argued that since the world is made up solely of matter and force, it follows that it must be uncreated and eternal. Contrary to some other materialists, he did not deny the existence of the human soul; but, to him, the soul was a kind of energy emerging from and depending on complex organic matter. With regard to the mind-body problem, he was a monist and not a dualist. The central theme in Büchner's materialist philosophy was the inseparability of force and matter. There could be no force without matter, and no matter without force. By conceiving force as bound to matter, Büchner countered the theists' exploitation of energy conservation for religious and spiritual purposes.

Materialism existed in many versions, some more radical than others. In Great Britain, scientists of a softened materialist inclination were known as scientific naturalists rather than materialists. They claimed that nature was nothing but a complex system of atoms and forces, but did not necessarily see science as antagonistic to religion. According to the leading physicist John Tyndall, a prominent scientific naturalist, the new view of science received solid support from the law of energy conservation. As he confidently stated in *Heat, a Mode of Motion*, a popular book on energy physics first published 1863, the law rigidly excludes both creation and annihilation.

Tyndall was referring to energy as well as matter. However, there was, at the time, a long tradition for considering the permanence of material atoms as an argument for God, who alone had the power to create them and make them disappear. Isaac Newton was happily unaware of the concept of energy, but he and his contemporaries made use of a similar argument related to the building blocks of matter. In Query 31 of his *Opticks*, Newton suggested that God — or what he called "an intelligent Agent" — had originally created the hard and solid particles called atoms: "And if he did so, it's unphilosophical to seek for any other Origin of the World, or to pretend that it might arise out of a Chaos by the mere Laws of Nature; though being once form'd, it may continue by those Laws for many Ages" (Newton, 1952, p. 402). Newton would not have agreed with Tyndall, and even less with Büchner.

1.1.2. MATTER AND ENERGY

Scientists in the second half of the nineteenth century had at their disposal two authoritative and fundamental laws. Apart from the new law of energy conservation, there was the much older law of mass conservation, which is usually ascribed to the great French chemist Antoine-Laurent Lavoisier, who formulated it as a general principle in the 1770s. The recognition that in a closed system chemical substances never disappear, but are only transformed into other substances with the same weight, was an important element in the chemical revolution during which modern chemistry was formed. Mass conservation quickly became accepted as a fundamental law that passed all experimental tests. By weighing the reactants in chemical processes with great precision before and after the reaction, in the early 1890s chemists proved that if there were a change in mass, it was unmeasurably small. They found the relative change $\Delta m/m$ to be less than 5×10^{-8} (Section 2.1.1).

The two conservation laws shared a paradigmatic status but were generally seen as separate, reflecting the ontological difference between matter and energy. Despite the revolutionary changes in twentieth-century physics, energy conservation still has an elevated, almost paradigmatic status, but this is only because the modern concept of energy has been redefined to include also mass. The fusion of the two concepts is embodied in the iconic formula $E = Mc^2$, which Einstein first stated as a consequence of his special theory of relativity in a brief paper of 1905. His formulation was this:

> "If a body gives off the energy L in the form of radiation, its mass diminishes by L/c^2. ... The mass of a body is a measure of its energy-content; if the energy changes by L, the mass changes in the same sense by $L/9 \times 10^{20}$, the energy being measured in ergs, and the mass in grams. It is not impossible that with bodies whose energy-content is variable to a high degree (e.g., with radium salts) the theory may be successfully put to the test" (Einstein *et al.*, 1952, p. 71).

To change the words into the modern standard formulation, we only have to replace the symbol L with M and, numerically, use the energy unit joule instead of the older erg unit ($1\,\text{erg} = 10^7\,\text{J}$).

With Einstein's theory, mass and energy merged into a single conserved entity. This stands out even more clearly if the equation is stated in so-called natural units (Section 1.4.1), where the value of the speed of light is set to $c = 1$. In that case, $E = M$ and energy is not just proportional to mass — in a formal sense, the two quantities are identical. Energy in the classical sense can be created or disappear, but only at the expense of an equivalent amount of mass. For example, when a positron and an electron annihilate, radiation energy is produced; and reversely, radiation energy may produce a positron–electron pair (see Section 1.3.3). Particle physicists rarely bother to include the factor $1/c^2$ and often state the mass of elementary particles in the energy unit eV, electron volt, say $m = 0.51\,\text{MeV}$ for the electron and $m = 938.3\,\text{MeV}$ for the proton ($\text{MeV} = 10^6\,\text{eV}$).

Although $E = Mc^2$ has a magical status in the eyes of the public, it is only one of several deductions from relativity theory. In the years after 1905, Einstein's theory caused much discussion and confusion among physicists, but the energy–mass equivalence formula was not seen as particularly shocking. Physicists were well prepared; for years before Einstein's theory, there had been suggestions of a somewhat similar kind. Some of them were qualitative speculations of little scientific value, such as Büchner's claim of 1855 that mass and energy are inseparable, whereas others were taken more seriously. For example, in a far-ranging address of 1886, the British chemist William Crookes imagined how the chemical elements might originally have come into being at some time in the far past. "Before this time matter, as we know it, was not," Crookes speculated (Crookes, 1886). "It is equally impossible to conceive of matter without energy, as of energy without matter."

About fifteen years later, many physicists endorsed the view that all matter consists of electrical charges generally called electrons and that these could be explained in terms of electromagnetic field theory without using the concept of mechanical mass. According to the "electromagnetic world view" popular at the time, the mass of an electron was partly or fully of electromagnetic nature and would therefore increase with the electron's velocity. Within this framework, electromagnetic energy was ascribed a mass, the two quantities being connected by an equation of the form $E \sim mc^2$. In the physics literature just preceding Einstein's 1905 paper, one can meet the equation $E = \frac{3}{4}mc^2$ derived from electromagnetic ether theory.

Yet another source for the belief in some kind of energy–mass relation was the perplexing phenomenon of radioactivity discovered by Henri Becquerel in 1896. Experiments in Paris showed that one gram of radium spontaneously generated an amazing 100 calories or 420 J per hour (Section 3.2.2). The source of the energy was a puzzle and would remain so for many years. As shown by the quotation from Einstein, he was aware that the radioactive energy might be due to a mass-to-energy transformation in accordance with $E = Mc^2$. Even before the advent of relativity theory, some scientists nourished similar thoughts. For example, in a book of 1904 the pioneer radiochemist Frederick Soddy clearly anticipated what Einstein had in mind:

> "It is not to be expected that the law of conservation of mass will hold true for radioactive phenomena. ... The total mass must be less after disintegration than before. On this view, atomic mass must be regarded as a function of the internal energy, and the dissipation of the latter in radioactivity occurs at the expense of the mass of the system" (Soddy, 1904, p. 180).

The fact that Einstein's energy–mass equation was anticipated by several scientists does not diminish its significance. First of all, Einstein's equation was exact and completely general, contrary to the suggestions made on the basis of electron theory, which presupposed matter to consist of electrical unit particles. The constitution of matter was of no relevance to Einstein's $E = Mc^2$, which would remain valid even if atoms and electrons did not exist. Although it took less than a decade before Einstein's equation entered established physics and advanced textbooks, for a long period of time it was not possible to verify it experimentally. It was only with the advent of experimental nuclear physics that the equation was indirectly tested and then with the expected result. In 1932, John Cockcroft and Ernest Walton studied a reaction where a beam of accelerated protons transformed a Li-7 target to two alpha particles:

$$^{1}_{1}\mathrm{H} + {}^{7}_{3}\mathrm{Li} \rightarrow {}^{4}_{2}\mathrm{He} + {}^{4}_{2}\mathrm{He} + 17.2\,\mathrm{MeV}.$$

They found that their results were consistent with $E = Mc^2$ to an accuracy of 5%. Much later experiments have reduced the deviation to ±0.00004%. In 1951, Cockcroft and Walton were awarded the Nobel Prize for their experiment, the first transmutation of atomic nuclei by artificially accelerated atomic particles.

Energy conservation in the generalized energy–matter version is considered to be strictly valid and absolutely fundamental, and yet it is neither true *a priori* nor a paradigm in the strong sense of the term. The law *can* be wrong, and throughout history there have been several attempts to introduce limited energy non-conservation. One example was the vexed problem of explaining the spectrum of beta rays from radioactive substances, which around 1930 seemed to contradict the conservation law. Niels Bohr seriously proposed that energy conservation is invalid within the atomic nucleus, an idea he defended for more than a year and thought might explain the energy produced by the stars (Section 3.2.4). As another example, this time on a cosmological scale, in the 1950s the steady-state theory of the universe assumed matter to be created continuously and spontaneously throughout space, albeit at the very small rate of 10^{-43} g/s/cm^3. The steady-state theory, more about which will follow in Section 4.1.4, turned out to be wrong, as did Bohr's earlier hypothesis of energy non-conservation.

1.1.3. THE SECOND LAW OF THERMODYNAMICS

When Mayer, Joule, and Helmholtz published their pioneering works on energy conservation, the law was not yet the first law of thermodynamics. In fact, the name "thermodynamics" first appeared in 1854, shortly after the law of energy conservation had been supplemented with a new, second law. Whereas the first law was static and about conservation, the second was dynamical and about evolution. What soon became known as the second law of thermodynamics had two fathers, Rudolf Clausius in Germany and William Thomson in Scotland. The latter, a towering figure in nineteenth-century physics, was knighted in 1892 as Baron Kelvin of Largs and is better known as lord Kelvin or just Kelvin.

In Clausius' first papers of 1850–1851, he stated the second law as a theorem of impossibility, namely that it is impossible to gain work by cooling matter below the temperature of the coldest of its surroundings. In a series of subsequent works, Clausius reformulated and generalized his theory, which in 1865 culminated with the introduction of the concept of entropy. The name was a neologism derived from a Greek word for transformation. Let the heat

Q of a system at absolute temperature T change by the infinitesimal amount dQ. Then, according to Clausius, the entropy difference between two states A and B of the system, $\Delta S = S_B - S_A$, is

$$\Delta S = \int_A^B \frac{dQ}{T},$$

where the path of integration corresponds to a reversible transformation from A to B. Armed with the entropy concept, Clausius famously stated the second law of thermodynamics as "the entropy of the universe tends towards a maximum," and likewise, for the first law, "the energy of the universe is constant" (Clausius, 1867, p. 365).

Thomson, who introduced the word "thermodynamics," never used the concept of entropy and rarely referred to it. Instead he formulated the second law as an inbuilt tendency in nature towards dissipation of energy. Imagine a situation in which two bodies at different temperatures are placed in contact, with heat being transferred from the warmer body to the colder without work being done. Energy is conserved, but as Thomson argued, it becomes dissipated in the sense that the system's capacity to perform work has diminished. In 1851, Thomson formulated his dissipation theorem as follows: "Any restoration of mechanical energy, without more than an equivalent of dissipation, is impossible in inanimate material processes, and is probably never effected by means of organized matter, either endowed with vegetable life or subjected to the will of an animated creature" (Thomson, 1852).

As appears from Thomson's formulation, he was not quite sure if the second law of thermodynamics was valid also for life processes. Indeed, whether formulated in terms of dissipation or entropy, for a long time it was a matter of debate if biological evolution processes might be exceptions to the second law. On the face of it, the evolution of life seems to contradict the tendency towards dissolution and disorder prescribed by the law, such as argued by several scientists and philosophers in the *fin de siècle* period. In 1910, the German physicist Felix Auerbach claimed that life processes are governed by an antientropic vital force called "ectropy," an idea which in modernized versions has survived to this day.

However, the pioneers of thermodynamics realized that the second law is valid only for an isolated system as a whole and not for individual parts of it. Local processes, such as the building up of complicated biomolecules from simpler ones, can occur even though they involve a decrease in entropy or energy dissipation. But an isolated living organism as a whole will inevitably decay.

Whereas the first law of thermodynamics had become part of chemical theory in the 1850s, the second law remained for a long time the exclusive domain of physics. When classical thermochemistry was finally replaced with chemical thermodynamics in the 1880s, the pioneers were two physicists, namely Josiah Willard Gibbs and Helmholtz (see also Section 2.4.1). In 1882, Helmholtz introduced entropy as concealed in what he called free energy and what is today known as the Helmholtz free energy $F = U - TS$, where U denotes the system's total energy. As Helmholtz pointed out, the value of F determines how a chemical reaction proceeds. By the turn of the century, chemistry had a solid foundation in thermodynamics, and yet many physical chemists found entropy to be an unwelcome concept, too abstract to be useful and to enter textbooks. Nernst's heat theorem, sometimes called the third law of thermodynamics, states in its modern version that entropy changes become zero at the absolute zero. However, when Walther Nernst formulated the theorem around 1910, he did not refer to entropy at all. He generally disliked entropy and wanted to avoid it.

The meaning and consequences of the second law of thermodynamics were hotly debated, not only by physicists and chemists but also by philosophers, theologians, and social critics. To the French philosopher Henri Bergson, it was "the most metaphysical of the laws of physics" (Bergson, 1965, p. 265).

Among the problems that attracted critical attention was that the second law, alone of all the fundamental laws of physics, refers to time and distinguishes between past and future. The influential Austrian philosopher-physicist Ernst Mach was not opposed to the second law, but he objected to Clausius' "scholastic" formulation that the entropy of the universe increases: "*If* the entropy of the world could be determined, it would be an absolute measure of time and it would be, at best, nothing but a tautology to say that the entropy of *the world* increases with time. Time, and the fact that certain changes take place in a definite sense, are one and the same thing" (Mach, 1923, p. 210). Mach referred to what came to be known as the entropic theory of time, the claim that entropy provides an objective measure of time. As Mach saw it, the statement that entropy increases with time is as meaningless as saying that time increases with time. Although Max Planck, a great authority on the second law of thermodynamics, disagreed with Mach in most respects, he shared his critical view with regard to the precise meaning of the law. According to Planck, "The energy and the entropy of the world have no meaning, because such quantities admit of no accurate definition. … There can be no talk of a definite maximum of the entropy of the world in any physical meaning" (Planck, 1911, p. 104).

The entropic theory of time criticized by Mach found a defender in the astronomer Arthur Eddington, who in 1928 coined the word "arrow of time" for a physical quantity discriminating between past and future. As he explained in a book a few years later, entropy was the only independent and reliable signpost for time: "Take an isolated system and measure its entropy S at two instants t_1 and t_2. We want to know whether t_1 is earlier or later than t_2 without employing the intuition of consciousness, which is too disreputable a witness to trust in mathematical physics. The rule is that the instant which corresponds to the greater entropy is the later" (Eddington, 1935, p. 55). As Eddington was well aware, the second law is not absolutely but only statistically valid. According to Ludwig Boltzmann's important gas theory of the 1870s, the essence of the second law is that a system, starting in an improbable state, moves spontaneously through a series of states that are ever more probable and finally ends in a maximum-probable state, or a state of maximum entropy. Boltzmann described the relationship by an equation, which in its modern notation reads

$$S = k_B \log W,$$

where k_B is Boltzmann's constant (1.38×10^{-23} J/K) and the quantity W is the number of microstates corresponding to a certain macrostate. The lower the value of W, the higher is the degree of ordering. The law of entropy increase therefore expresses a spontaneous increase in disorder.

It follows from Boltzmann's probabilistic interpretation that there is a non-zero probability that the entropy of a closed system decreases, although the probability is so small that it can be ignored in almost all cases. Boltzmann calculated the time it would take for the 10^{19} molecules in 1 cm^3 of a gas to return to their initial state and found an inconceivably large number of years. Still, if we do something sufficiently many times, even the most extraordinary things will happen. Strictly speaking, they are not impossible, only highly unlikely. If we put a kettle of water on a hotplate and turn on the current, the water will get hot and eventually boil. But if we do it again and again an incredible number of times, we will experience an instance when the water will freeze miraculously to ice. Eddington illustrated the situation with a limerick:

> "There once was a brainy baboon
> Who always breathed down a bassoon
> For he said, It appears
> That in billions of years
> I shall certainly hit on a tune" (Eddington, 1935, p. 62).

As far as Eddington was concerned, one could forget about antientropic processes and the strange consequence of time going backwards apparently associated with such processes.

1.1.4. HEAT DEATH AND ENTROPIC CREATION

Of the many problematic features of the second law of thermodynamics, the so-called "heat death" was the most alarming and the one which made the concept of entropy highly controversial. Assuming the second law to be valid for the universe at large, it follows that in the far future, when the entropy is approaching its maximum, there will be no temperature difference between any two positions in space. All order and organization in the universe will be lost, with no possibility of regaining its former life and glory. The heat death scenario was first enunciated by Helmholtz in a popular lecture of 1854, and soon afterwards it was restated by authorities such as Clausius and Thomson. In 1868, Clausius phrased the prediction of a heat death in terms of his new concept of entropy: "The more the universe approaches this limiting condition in which the entropy is a maximum, the more do the occasions of further change diminish; and supposing this condition to be at last completely attained, no further change could evermore take place, and the universe would be in a state of unchanging death" (Clausius, 1868, p. 419).

As Clausius pointed out, the second law contradicted the notion of a cyclic or recurrent universe, the popular view that in the long run the overall state of the world remains the same, with decay processes in some parts of the universe being counterbalanced by growth processes in other parts. The heat death scenario created massive attention not only among scientists but also in the public arena. Could it really be true that life would once cease never to return? Charles Darwin found the heat death disturbing because it was so clearly incompatible with his optimistic belief in greater perfection obtained by progressive evolution. In his autobiography, he referred to the physicists' heat death, finding it "an intolerable thought that he [man] and all other sentient beings are doomed to complete annihilation after such long-continued slow progress" (Darwin, 1958, p. 67).

The half-century between 1865 and 1915 witnessed a variety of suggestions to avoid what Clausius and his allies claimed was unavoidable. Many of the period's antiheat-death proposals appealed to hypothetical counter-entropic mechanisms, such as collisions between stars and other celestial bodies. A solution of this kind was developed in detail by the Swedish physical chemist Svante Arrhenius, a Nobel Prize laureate of 1903, who speculated that the radiation pressure arising from collisions between stars and nebulae would keep the present world system going for an indefinite period of time. Other scientists argued that the entropy law was invalid for the supposedly infinite universe, and others again suggested that the entropy would never reach a maximum but only increase asymptotically and thus allow eternal cosmic evolution. Most of the attempts to avoid the heat death built on ad hoc assumptions and wishful thinking, and they did not seriously disturb the consensus view of experts in thermodynamics, namely that the heat death was an inescapable consequence of the second law. Outside science, some commentators simply dismissed the law with the argument that if it led to the absurdity of a dead universe, the law surely must be wrong.

The heat death scenario was about the state of the universe in the very far future, but it could easily be turned around and presented as an argument for a beginning of the universe a finite time ago. What has been called the entropic creation argument was first developed in the late 1860s, and it can be formulated as a simple syllogism consisting of two premises and one conclusion:

(1) The entropy of the world increases continually
(2) Our present world is not in a state of very high entropy
(3) Hence, the world must be of finite age.

Of the two premises, (2) is an indisputable empirical fact, since a world of very high entropy would be entirely different from the one we know. Premise (1) is Clausius' version of the second law, assuming that the law can be applied to the universe *in toto*. This straightforward syllogism was one of the first scientifically based arguments for an origin of the universe, and as such it of considerable historical interest. When it became popular as well as controversial in the nineteenth century, it was because it was discussed in an extended and apologetic form, as an argument for the existence of God. The logic was that if the universe had a beginning, it must have been created, and the creator must be the omnipotent God. Peter Guthrie Tait, a respected British mathematical physicist and collaborator of Thomson, gave voice to the entropic creation argument in a book of 1871. Speaking of the law of energy dissipation rather than the entropy law, he said:

> "As it alone is able to lead us, by sure steps of deductive reasoning, to the necessary future of the universe — necessary, that is, if physical laws remain forever unchanged — so it enables us distinctly to say that the present order of things has *not* been evolved through infinite past time by the agency of laws now at work; but must have had a distinctive beginning, a state beyond which we are totally unable to penetrate; a state, in fact, which must have been produced by other than the now visibly acting causes" (Tait, 1876, p. 17).

Tait's creative cause was of course God.

Although the cosmological consequences of the second law were stated in terms of "beginning" and "end," this should not be understood literally, in the meaning of cosmic creation and annihilation in the strong sense. Such a meaning — where the space, energy and matter of the universe appears from nothing and disappears into nothing — makes sense within the framework of relativistic cosmology with its big bang and big crunch (Chapters 4.1 and 4.3), but it was outside the mental framework of scientists in the Victorian era. The most the entropic argument could do and can do is to lead to a beginning of changes and activity in the universe, just as the heat death is an end of changes and activity, not of the universe itself. When the universe had reached its end at the maximum entropy state, it would still exist, but in a frozen or dead form where time no longer has any meaning.

The heat death scenario of the nineteenth century has not been called off, although in modern science it has a status different from the one in the Victorian era. Recent developments in physics and cosmology have not led to a clarification of the problem of cosmic entropy, only made it more complex. After all, Clausius, Thomson, and their contemporaries were happily unaware of the gravitational field of general relativity and of black holes, dark matter, and dark energy. Most realistic models of the universe predict a kind of heat death, but it is doubtful if the universe as a whole can meaningfully be ascribed a definite content of entropy. The modern view is thus in line with the one argued much earlier by Mach and Planck, both of whom denied that the entropy of the world is a well-defined quantity. Nonetheless, modern extrapolations of the history of the universe to the very distant future — what is sometimes called "physical eschatology" — lead to scenarios no less pessimistic than those discussed in the late nineteenth century.

1.1.5. CULTURAL AND IDEOLOGICAL IMPACTS

Whether they were directly inspired by the thermodynamic heat death or not, several poets in the Victorian era took up the theme of a dying world. Here is one verse from 1866, composed by the British poet Algernon Swinburne:

> "Then star nor sun shall waken,
> Nor any change of light:
> Nor sound of waters shaken,
> Nor any sound or sight:
> Nor wintry leaves nor vernal,
> Nor days nor things diurnal.
> Only the sleep eternal,
> In an eternal night" (Kragh, 2008, p. 38).

The heat death and its associated argument for a beginning of the universe gave rise to an extensive debate in late nineteenth-century science and culture. In this controversy, scientific arguments were often secondary to, or mixed up with, beliefs (or disbeliefs) of a religious, social and ideological nature. With few exceptions, conservatives and religiously oriented authors tended to accept the heat death, which after all had a superficial similarity with the apocalyptic passages in the Bible. On the other hand, materialists, atheists, and positivists were convinced that the universe is a self-generating *perpetuum mobile,* and they consequently denied the heat death and the idea of a cosmic beginning. Moreover, this conviction went hand in hand with a dogmatic belief in the infinity of the universe, which they conceived as an infinite number of stars and galaxies.

The discussion of the wider implications of thermodynamics was particularly visible in the cultural struggle that unfolded in Otto von Bismarck's new Germany and is known as the *Kulturkampf*. Many Jesuits and other Catholics entered the struggle by arguing that their faith was in harmony with science, and as part of this strategy they studied critically and in detail the relationship between thermodynamics and Christian belief. However, most of them resisted the temptation to use thermodynamics apologetically, as a scientific proof of God. They were pleased to note that the second law supported theism, but most Catholic authors concluded that thermodynamics failed in demonstrating with certainty that the world must have a beginning and an end. The discussion cast long shadows, and as late as

1951 it resurfaced in a controversial so-called encyclical address given by pope Pius XII to the Pontifical Academy of Sciences in Rome. As the pope made clear, he considered the heat death to be an additional argument for a universe subordinated to the will of God:

> With the law of entropy, discovered by Rudolf Clausius, it became known that the spontaneous processes of nature are always related to a diminution of the free and utilizable energy, which in a closed material system must finally lead to a cessation of the processes on the macroscopic scale. This fatal destiny, which … comes from positive scientific experience, postulates eloquently the existence of a Necessary Being.[1]

Many contributors to the entropic discussion argued, or rather took it for granted, that the second law of thermodynamics could only be applied to a finite universe. If the universe were infinite, such as claimed by materialist thinkers, there was no longer any scientific justification for the heat death and a cosmic beginning. In reality, the finite universe was a dogma among most conservatives and theists, and the infinite universe was no less of a dogma among materialists and atheists. There were a few exceptions, such as Thomson who was a sincere Christian and had no doubts about the heat death as a consequence of the second law, but nonetheless found it impossible to believe in a finite universe. From a more modern point of view, the question of an infinite universe is dealt with in Chapter 4.2.

To illustrate the ideological and political dimension of the entropic-cosmological debate, consider Karl Marx's loyal friend and collaborator Friedrich Engels, a key figure in early communist thought. Engels, who had a broad interest in science, was early on acquainted with the two laws of thermodynamics, including Clausius' formulation of them. He detested the law of entropy increase, not so much for scientific reasons but because he found it to be politically incorrect and ideologically dangerous. In a remarkable letter to Marx of 21 March 1869, he expressed his misgivings

[1] http://www.academyofsciences.va/content/accademia/en/magisterium/piusxii/22november1951.html.

about the second law and its unpalatable idea of a fundamental irreversibility in nature. Engels described what he called the "very absurd theory" as follows: "That the world is becoming steadily colder, that the temperature of the universe is leveling down and that, in the end, a moment will come when all life will be impossible and the entire world will consist of frozen spheres rotating round one another."[2] And with respect to the ideological problem inherent in the second law, he continued:

> I am simply waiting for the moment when the clerics seize upon this theory [Clausius'] as the last word in materialism. It is impossible to imagine anything more stupid. Since, according to this theory, in the existing world, more heat must always be converted into other energy that can be obtained by converting other energy into heat, so the original hot state, out of which things have cooled, is obviously inexplicable, even contradictory, and thus presumes a God.

Engels was deeply committed to the idea of an infinite, eternally recurrent universe, which he regarded as an integral part of atheism and dialectical materialism. When Marxism–Leninism became the state philosophy of the Soviet Union, and later also of Red China, Engels' view of dialectical materialism, including his insistence of an infinite universe, received official approval. As a consequence, cosmological ideas of a finite universe became for a period *theoria non grata*.

[2] http://www.marxists.org/archive/marx/index.htm.

.1.2

The Versatile Bohr Atom

1.2

The Versatile Bohr Atom

Planck's quantum of action h was first applied to the structure of atoms in 1910, but it was only with Niels Bohr's path-breaking theory three years later that quantum theory truly became the basis of atomic and molecular physics. Bohr's theory built on a non-classical foundation and, for this reason, it evoked strong opposition in parts of the physics community. Nonetheless, its undeniable explanatory and predictive strength made it appealing and even necessary. By 1915, it was accepted by many avant-garde physicists. The consequences of some of the early predictions of Bohr's atomic theory, such as the spectroscopic isotope effect and the existence of so-called Rydberg atoms, were only recognized many years later and are today thriving research areas. Bohr's theory was as much aimed at chemistry as at physics, but in this respect it was less successful, if far from unimportant. In particular, it resulted in an atomic explanation of the full periodic system of the elements that essentially was in agreement with the currently accepted theory.

1.2.1. A QUANTUM THEORY OF ATOMIC AND MOLECULAR STRUCTURE

In the fall of 1913, Niels Bohr published in *Philosophical Magazine* a series of three papers with the common title "On the Constitution of Atoms and Molecules." What is known as Bohr's trilogy is rightly considered a watershed in atomic theory and physics generally. From a conceptual point of view, his new and very ambitious theory of the constitution of atoms and molecules was based on two postulates or "principal assumptions," as he called them.

Both postulates were of a radical nature and could only be indirectly justified by the consequences they led to in terms of experimentally testable results. According to the first postulate, atoms could only exist in certain "stationary states" characterized by discrete energies, or discrete radii of the revolving electrons, and labelled by a single quantum number n. The lowest energy state or ground state $n = 1$ was the one in which the atom would exist under normal circumstances. Whether in the ground state or in some of the higher excited states, ordinary mechanics was assumed to be valid whereas electrodynamics was claimed to be invalid. The second postulate was concerned with the apparently spontaneous transition of an atomic system from one stationary state $E(n_2)$ to another state $E(n_1)$, stating that the process would be followed by the emission or absorption of homogeneous radiation in the form of a quantum of energy. "The frequency and the amount of energy emitted is the one given by Planck's theory," Bohr stated in his 1913 paper. That is, the frequency ν of the radiation was given by the energy difference

$$E(n_2) - E(n_1) = h\nu,$$

where h is Planck's constant.

Two points are worth noticing. First, and contrary to what is often stated in textbooks, the quantum transitions or "jumps" in Bohr's atoms did *not* involve light quanta or photons but monochromatic electromagnetic waves. Bohr was at the time aware of Einstein's controversial hypothesis related to the photoelectric effect, which he referred to in the first part of his trilogy, but he did not believe in the light quantum at all. Indeed, for more than a decade he remained a staunch opponent of Einstein's light quanta or what soon became known as photons (the name "photon" was first suggested in 1916, but in a sense that differed substantially from Einstein's light quantum). In his Nobel Prize lecture given in Stockholm in 1922, Bohr admitted that Einstein's hypothesis was of "heuristic value," but then emphasized that it "is quite irreconcilable with so-called interference phenomena [and] is not able to throw light on the nature of radiation" (Bohr, 1923).

Second, while in Bohr's original theory the stationary states were ascribed a sharply defined energy ($\Delta E = 0$), in 1918 and later works he argued from the correspondence principle that the energy of a stationary state would vary over a small interval. Instead of a definite E, the energy would be given by $E \pm \Delta E$, with ΔE much smaller than E. Bohr realized that with this modification discrete spectral lines must have a natural line width even at absolute zero temperature, where the Doppler broadening due to the motion of the light-emitting atoms is absent. After the introduction of quantum mechanics, the impossibility of spectral lines with zero broadening was shown to be a consequence of Heisenberg's uncertainty principle in the form of $\Delta E \Delta t \geq \hbar$.

Bohr's theory was initially received with reservation or, in some quarters, hostility. Physicists in England and Germany generally rejected, or at least found unattractive and bewildering, the theory's conceptual basis such as given by the two postulates. They found it hard to swallow a theory which claimed that established knowledge of physics failed on a subatomic scale, and that too without offering a physical mechanism explaining the mysterious quantum transitions. Nonetheless, within a short span of years, many mainstream physicists accepted the theory and began working on it. The primary and almost sole reason for the acceptance was the theory's amazing explanatory and predictive strength, its ability to account quantitatively for known facts and predict new ones that turned out to agree beautifully with measurements.

The success was most conspicuous within the realm of optical spectroscopy, if not limited to this branch of science. This was what impressed Robert Millikan, the eminent American experimentalist. At the age of 83, Millikan described how Bohr's theory had revolutionized spectroscopic studies. "The immense field of spectroscopy was essentially an unexplored dark continent prior to the advent of Bohr's theory," he stated (Millikan, 1951, p. 110). "Bohr's equation has been the gateway through which hundreds of explorers have since passed into that continent until it has now become amazingly well mapped." Interestingly, Bohr had originally no interest in spectroscopy and developed his first version of the atomic theory without considering the emission and absorption of light. As late as 31 January 1913, he wrote to Ernest Rutherford, "I do not at all deal with the question of calculation of the frequencies corresponding to the lines in the visible spectrum" (Kragh, 2012, p. 56). Only a month or so later did Bohr realize the crucial significance of Balmer's spectral formula and spectroscopy generally.

Based on his atomic model and the frequency condition $\Delta E = h\nu$, Bohr showed that a one-electron atom passing from state n_2 to n_1 ($n_2 > n_1$) would give rise to the frequencies

$$\nu = Z^2 R_H c \left(\frac{1}{n_1^2} - \frac{1}{n_2^2} \right),$$

where Z denotes the nuclear charge and R_H is the empirically known Rydberg constant for hydrogen. Thus, not only did he derive a generalized version of the Swiss schoolteacher Johann Balmer's formula known since 1885 ($n_1 = 2$), he also explained the Rydberg constant in terms of fundamental constants of nature. With m denoting the electron's mass, he found

$$R_H = \frac{2\pi^2 m e^4}{h^3 c} = 109\,740\,\mathrm{cm}^{-1},$$

in convincing agreement with the measured value $R = 109\,675$ cm^{-1}. The predicted hydrogen lines in the infrared ($n_1 > 3$) and extreme ultraviolet ($n_1 = 1$) regions were not known in 1913, but when they were detected during the following decade they turned out to agree perfectly with Bohr's expression. The same was the case with hydrogen's ionization energy, which according to Bohr must be the energy difference between the states $n_2 = \infty$ and $n_1 = 1$ given by

$$E_{ion} = R_H ch = \frac{2\pi^2 m e^4}{h^2} \cong 13\,\mathrm{eV}.$$

At the time, hydrogen's ionization energy was only approximately known from experiments, which suggested a value of about 11 eV, but as later experiments proved, Bohr's predicted value was more precise. Bohr also found an expression for the radius a_n of a hydrogen atom depending on its quantum state, namely

$$a_n = n^2 \frac{h^2}{4\pi^2 me^4}.$$

For the ground state $n = 1$, the numerical value given by Bohr was 0.55×10^{-8} cm. The present value for what is known as the Bohr radius is 0.529×10^{-8} cm.

Even more impressive was Bohr's response to what was widely thought to be a new hydrogen series first detected in 1896 by the American astronomer Edward Pickering. The new series was apparently proven in laboratory experiments from 1912 made by the British astrophysicist Alfred Fowler, who in a hydrogen-helium mixture found the line $\lambda = 4686$ Å. This line also appeared in Pickering's stellar spectra. According to Fowler, the frequencies of the new hydrogen series could be expressed by a Balmer-like formula with half-integral denominators, namely of the form

$$\nu = R_H c\left(\frac{1}{2^2} - \frac{1}{(n+½)^2}\right).$$

The prominent 4686 line corresponded to $n = 3$. The Pickering-Fowler hydrogen lines were obviously a severe problem for Bohr's theory, which only allowed integral denominators corresponding to the principal quantum number n. Indeed, were the lines really due to hydrogen, the theory was wrong. However, Bohr brilliantly escaped the threat by simply rewriting the empirical expression as

$$\nu = 4R_H c\left(\frac{1}{4^2} - \frac{1}{(2n+1)^2}\right).$$

This showed that the lines must be attributed to ionized helium (He^+ with $Z = 2$) and not to hydrogen. Experiments made in Rutherford's laboratory in Manchester with pure helium confirmed Bohr's hypothesis and thus provided his atomic theory with additional evidence. Much later, the Hungarian philosopher Imre Lakatos commented on this episode in the history of atomic physics. "Thus the first apparent defeat of [Bohr's] research programme was turned into a resounding victory," he wrote (Lakatos, 1970, p. 149). To Lakatos, it was a beautiful case of what he called "monster-adjustment," by which he meant "turning a counter-example, in the light of some new theory, into an example." Bohr repeated his monster adjustment in the case of data from other stellar spectra that indicated hydrogen frequencies corresponding to denominators of the kind $(n \pm 1/3)^2$. In this case, he suggested that the lines were due to the hydrogen-like Li^{2+} ion with $Z = 3$.

Fowler reluctantly acknowledged Bohr's explanation, but without accepting the theory and still dissatisfied with the accuracy of its predictions. As he pointed out in the autumn of 1913, Bohr's theoretical values for the wavelengths of H and He^+ differed slightly but significantly from precise observations. The new challenge did not bother Bohr too much. He just removed the simplifying assumption of the electron revolving around a fixed nucleus of mass M. In reality, the two particles would move around the common center of gravity, which could easily be accounted for by replacing the electron mass m with the reduced mass μ given by

$$\mu = \frac{mM}{m+M} = \frac{m}{1+m/M}.$$

In this way, the discrepancies between theory and experiment disappeared. Moreover, it followed that the Rydberg constant for a very heavy nucleus differed from R_H according to

$$R_\infty = R_H\left(1+\frac{m}{M_H}\right).$$

Bohr reported $R_\infty = 109735\,\text{cm}^{-1}$, whereas the modern value is $1097357\,\text{cm}^{-1}$. In a lecture given in 1914, Fowler used Bohr's expressions to derive a value for the mass ratio of the hydrogen nucleus and the electron, which is today recognized as one of the important dimensionless constants of nature. Fowler's result $M_H/m = 1836 \pm 12$ agrees with modern data.

Although the early successes of Bohr's theory were predominantly related to spectroscopy, the theory also resulted in other noteworthy insights. For example, in the second part of the trilogy Bohr argued that beta radioactivity had its origin in the nucleus and not, as believed by Rutherford and most other physicists, in the surrounding sphere of electrons. As Bohr pointed out at a time when the term "isotope" had not yet been coined, it was known that some radioactive isotopes belonging to the same element emitted beta rays of different energies. On the new assumption that isotopes differed in mass but not in their electron system, this meant that beta rays must come from the atomic nucleus just like the alpha rays. By the end of 1913, Bohr's view of the origin of beta rays was adopted by Rutherford, Frederick Soddy, and most other experts in radioactivity.

1.2.2. MISUNDERSTANDINGS AND CONCEPTUAL OBJECTIONS

Bohr readily admitted that his new theory of atomic structure was a strange mixture of classical physics and radical departures from it based on the two quantum postulates. He knew that the theory contained several arbitrary and unsatisfactory features, but these he considered to be necessary for the time being. As he repeatedly stressed, the theory he proposed in 1913 was a work in progress and not a final product. Other physicists looked at it with different eyes and, if interested in the theory at all, generally preferred to focus on its empirical results rather than its unpalatable basic assumptions. For example, the British physicist and astronomer James Jeans was sympathetic to the theory at an early stage. He found its account of the spectra to be ingenious and suggestive, but at the same time he objected that "the only justification at present ... for these [basic] assumptions is the very weighty one of success" (Jeans, 1913).

The Dutch physicist Paul Ehrenfest was thoroughly familiar with quantum theory and eventually became an enthusiastic advocate of Bohr's atomic theory. But, initially, he thought it was nothing less than "monstrous" and "horrible," such as he phrased it in a letter to Arnold Sommerfeld of 1916. The most persistent, and the most formidable, of Bohr's opponents was the British mathematical physicist John Nicholson, who in 1911 had introduced Planck's quantum of action in his own nuclear theory of atomic structure. From 1913 to about 1918, Nicholson campaigned against Bohr's model, which he examined in minute, intricate, technical detail. He argued that Bohr's models for many-electron atoms were mechanically unstable and that even his model for the simple He^+ ion was wrong. Although Bohr chose to ignore Nicholson's criticism, for a while it was taken seriously by British physicists and astrophysicists.

What bothered many physicists can be summarized in four questions. First, with what right did Bohr declare the authoritative theory of electrodynamics to be invalid in the stationary states? Second, how could radiation with a certain frequency be emitted if it did not correspond to the frequency of vibrating or rotating electrons in the atom? Third, the discontinuous quantum jumps were completely mysterious, for how could an electron know where to end its jump from a higher excited state? Fourth, how could an electron possibly pass from a higher to a lower state if the space in between was strictly forbidden for it? There was much to worry about.

Many physicists complained that although Bohr had provided a formal scheme that admirably reproduced known and unknown spectral lines, he had not *explained* the lines in terms of a model with a definite mechanism. In J. J. Thomson's classical atomic model, the emitted frequencies had their origin in the frequencies of electrons rotating and vibrating in a positively charged fluid, but there was no similar mechanism in Bohr's theory. On the contrary, here the mechanical frequencies bore no relation to the radiation frequencies. The only analogy with electrodynamics was for transitions between very high quantum states, where the orbital motion is slow. In this case, Bohr proved with an early correspondence argument that the emitted frequencies were multiples of the orbital or mechanical frequency.

The belief that the frequency of emitted light must correspond to something actually vibrating in the atom was deep-seated and was a major reason why many physicists were unhappy with Bohr's theory. Einstein, no less a radical mind than Bohr, was, in this respect, an exception for he had no problem at all with the lack of pre-existing frequencies in Bohr's theory. On the contrary, in a conversation of late September 1913 with the Hungarian chemist George Hevesy, a close friend of Bohr, he singled out this feature as a reason for believing in the essential truth of the theory. As Hevesy reported in letters to Bohr and Rutherford, Einstein called the new theory "an enormous achievement" and "one of the greatest discoveries."

To most other physicists, the lack of consistency with the Maxwell–Lorentz theory of electromagnetism remained a serious problem requiring some fundamental modification of the atomic theory or perhaps a revision of electrodynamics in the interior of atoms. In the period from about 1914 to 1920 there were several attempts to interpret the quantum conditions on a classical basis, either from electrodynamics or hydrodynamics, but none proved viable. Tellingly, the Nobel physics committee turned down Bohr's candidacy for the 1919 prize, with the argument that Bohr's ideas were "in conflict with physical laws that have so long been considered indispensable ... [and] therefore cannot be considered to contain the real solution to the problem" (Pais, 1991, p. 214).

The German experimentalist Johannes Stark, a Nobel laureate of 1919, had a strong dislike of Bohr's theory, which he thought was empirically inadequate and philosophically absurd. In a paper published in 1917, he put his finger on the theory's problem with causality: "How can an electron, when it is leaving its initial orbit, already know its final orbit; or, by which considerations does it choose its final orbit, so that already before it has attained it [knows] the definite frequency it has to emit to reach the chosen final orbit?" (Stark, 1917). It was a most sensible question, and neither the first nor the last time it was asked. As early as March 1913, in a letter commenting on Bohr's first draft of his treatise, Rutherford addressed the very same question, which he considered "a great difficulty." Bohr did not consider the question to be important or even relevant, and his answers to it were characteristically indirect and evasive. He stressed that his theory was of a formal nature and not concerned with the radiation mechanism and that it did not allow a detailed space-time description of atomic processes.

Nonetheless, the question continued to be raised and discussed by several physicists other than Rutherford and Stark. In some way or another, it seemed that an electron in an excited state "knew" in advance the lower state to which it would transit. Explanations in physics are causal and not teleological, and yet Bohr's atomic electrons apparently behaved as if they were governed by a final purpose rather than a preceding cause. Did the quantum electron possess a free will or something analogous to it? The question was raised by more than a few physicists, and only half-jokingly. In a popular book of 1923, Bohr's close collaborator, the Dutchman Hendrik Kramers, explained in plain language the odd behavior of a Bohr electron: "Even from the very beginning the electron seems to arrange its conduct according to the goal of its motion and also according to future events. But such a gift is wont to be the privilege of thinking beings that can anticipate certain future occurrences" (Kramers and Holst, 1923, p. 137).

Of course, Bohr did not provide his electrons with free will, but neither did he believe that the emission of radiation was a causal process in the sense of classical physics. He assumed that somehow the process depended on both the initial and the final state and that it occurred spontaneously, only governed by a law of probability. Einstein was far from foreign to probabilistic radiation laws, but he resisted the lack of causality that appeared in Bohr's theory of emission and absorption of light. By 1924, Bohr had developed his theory into a more advanced version that still retained the fundamental features of the original theory. In a letter to Max Born, Einstein expressed his worries about Bohr's new radiation theory: "I find the idea quite intolerable that an electron exposed to radiation should choose *of its own free will*, not only its moment to jump off, but also its direction. In that case, I would rather be a cobbler, or even an employee in a gaming-house, than a physicist" (Born, 1971, p. 82). Einstein's remark anticipated the heated discussion concerning quantum jumps that would soon enter quantum mechanics and has continued up to this day.

If the fundamental law of causality were at stake with Bohr's theory, so was, according to some critics, the equally fundamental law of energy conservation. One of the critics was the American physical chemist Gilbert N. Lewis, who had his own and very different ideas of the constitution of atoms and molecules. Lewis thought that the postulate of stationary states was logically inconsistent and that the concept of a ground state was deeply problematical, since the electron continued to revolve at great speed even at zero absolute temperature. This he considered to be contrary to the fundamental law of energy conservation. In *Valence*, a widely read book of 1923, Lewis invented a thought experiment in which an atomic electron in its ground state induced an alternating current in a piece of metallic wire close to the atom. The current would generate ohmic heat in the wire, and yet there was no source for the heat energy; after all, the electron would just continue to rotate at the same speed, maintaining its kinetic and potential energy. According to Lewis, Bohr's atomic electron moved in such a way that it produced no physical effect whatsoever, and this he found to be nonsensical. He thought that the paradoxes inherent in Bohr's atomic theory somehow reflected its basis in what he called "scientific bolshevism," namely quantum theory. The connection between bolshevism and quantum theory is not obvious, but they were both revolutionary ideas threatening the established order, in one case of society and in the other of science.

Bohr did not respond to Lewis' objections. As far as the charge of energy non-conservation is concerned, it would not have frightened Bohr, who did not consider *strict* energy conservation to be a sacrosanct principle (see Section 3.2.4). In a manuscript of about 1917, he referred cautiously to the possibility of energy non-conservation, a concept that appeared explicitly in his later theory of light-matter interaction known as the BKS or Bohr–Kramers–Slater theory. In this theory, energy was not conserved for individual atomic processes. Einstein's comment on the electron's free will was related to the BKS theory in particular.

1.2.3. MONSTER ATOMS

The classical Daltonian atom as adopted by most chemists and physicists in the nineteenth century was an immutable and indivisible tiny sphere, the size of which depended only on the chemical element of which it was the sole constituent. By 1910, the atom was no longer indivisible but it still had a definite size, a radius of the order 10^{-8} cm. Bohr disagreed, for according to his theory the size of an atom was not a fixed quantity. In the case of the hydrogen atom, he showed that the radius a depended on the state of excitation as

$$a = n^2 a_1,$$

where n is the quantum number and a_1 the Bohr radius. Bohr emphasized this unusual feature of his theory, namely that in principle there was no upper limit to the size of an atom. In the first part of his 1913 paper, he remarked that for $n = 33$, "the diameter [of a hydrogen atom] is equal to 1.2×10^{-5} cm, corresponding to the mean distance between the molecules at a pressure of about 0.02 mm mercury" (Bohr, 1913). Since the volume of an atom varies as a^3 or n^6, for $n = 33$ it is more than one billion times the volume of the ordinary atom in its ground state!

What Bohr may not have known is that the idea of such monster atoms can be found in the earliest version of atomism credited to the ancient Greek philosopher Democritus, a contemporary of Socrates. Whether or not Democritus really believed in macroscopic atoms, this is how the later commentator Diogenes Aetius, a Greek living in the fourth century AD, interpreted him. According to Aetius, "Democritus ... claims that it is possible for an atom to exist as large as a world" (O'Brien, 1981, p. 223). Bohr did not go quite as far as Democritus, but a hydrogen atom of radius 10^{-5} cm or even more was still a remarkable claim.

The highly excited atoms should give rise to spectral lines corresponding to the large quantum numbers, say from $n = 30$ to $n = 2$. However, no such lines were detected in laboratory experiments with vacuum tubes, which revealed only 12 Balmer lines. As Bohr pointed out, astronomers had observed as many as 33 lines belonging to the Balmer series. How could it be that observations of stars and nebulae were, in this respect, superior to laboratory experiments? Bohr suggested that it was readily understandable according to his theory. Spectral lines arising from hydrogen in high quantum states required a density much lower than that attainable in vacuum tubes, and also, to obtain a sufficiently intense radiation, a very large space filled with hydrogen gas. "If the theory is right," Bohr wrote in his 1913 paper, "we may therefore never expect to be able in experiments with vacuum tubes to observe the lines corresponding to high numbers of the

Balmer series of the emission spectrum of hydrogen." On the other hand, the conditions existed in the rarefied atmospheres of stars and nebulae, which explained the astronomers' success.

Since about 1970, the kind of very large atoms introduced by Bohr have been known as "Rydberg atoms" because the frequencies of such atoms were included in a hypothetical spectral formula that Johannes Rydberg proposed in 1890. However, the Swedish physicist did not attempt to explain the lines of his formal scheme, nor did he relate them to particular states of atoms. The name is thus unfortunate, while "Rydberg-Bohr atoms" seems a more appropriate name. In any case, Bohr's prediction of highly excited atoms in outer space was eventually vindicated, if only after his death and not in the optical region.

Because of the extremely low density in interstellar gas clouds, in these environments Rydberg atoms can exist for comparatively long periods of time without being ionized. Whereas the lifetime of an ordinary excited atom is of the order 10^{-8} second, Rydberg atoms in space may live as long as a second. In the summer of 1965, Bertil Höglund and Peter G. Mezger from NRAO, the National Radio Astronomy Observatory established in 1957, searched for radiation generated by electron–proton recombination in the interstellar medium. They serendipitously detected microwave radiation from hydrogen atoms in the Omega and Orion nebulae (M17, M42), which corresponded to the transition from $n = 110$ to $n = 109$. (The quantity n is the principal quantum number in quantum mechanics, one unit lower than n in Bohr's old quantum theory). Later, radio astronomers found states as high as $n = 350$ in outer space, and in 2014 Leah Morabito and collaborators at Leiden University detected carbon atoms in quantum state 508 in the starburst galaxy M82.

Today, the study of Rydberg atoms has grown into a minor industry, with numerous applications beyond astrophysics. With the development of tunable dye lasers in the 1970s, it became possible to study Rydberg atoms in the laboratory, which stimulated new research ranging from fundamental physics to quantum computing and military technologies. Even though the study of Rydberg atoms is, today, an important part of the physical sciences, few researchers are aware that it all started with Bohr's pioneering analysis of 1913. Even fewer are aware of Democritus' possible anticipation.

1.2.4. THE ISOTOPE EFFECT

There were more gems hidden in Bohr's treatise on atomic structure. In the second part of the trilogy, Bohr referred to isotopes without using the name, and in an interview conducted shortly before his death in November 1962, he even claimed that he was the first who got the idea of isotopy. Perhaps he got it independently, but it is generally recognized that the priority belongs to Soddy, who is also credited with coining the word "isotope." Soddy was indeed the first to use the name, but it was actually coined by the physician and author Margaret Todd, who suggested it to Soddy in a conversation with him. Todd, who was proficient in Greek, knew that "same place" (in the periodic system) could be transcribed to "isotope."

Having introduced the reduced mass in his model of the hydrogen atom, it occurred to Bohr that the same reasoning could be applied to the spectral lines of isotopes. Their different masses would, at least theoretically, result in a small spectral shift. In late 1913, Bohr speculated that J. J. Thomson's recent evidence for the unusual triatomic hydrogen molecule H_3 (Section 2.3.1) might instead be interpreted in terms of a hydrogen isotope of atomic weight 3, what came to be known as tritium, symbol T. If this isotope existed, it would yield a shift in wavelength relative to ordinary hydrogen equal to

$$\lambda_{\mathrm{T}} - \lambda_{\mathrm{H}} = \left(1 - \frac{R_{\mathrm{H}}}{R_{\mathrm{T}}}\right)\lambda_{\mathrm{H}} = \frac{2m}{3m + 3M_{\mathrm{H}}}\lambda_{\mathrm{H}} = 3.6 \times 10^{-4}\,\lambda_{\mathrm{H}}.$$

Bohr believed that the shift might be measurable and, together with the Copenhagen physicist Hans M. Hansen, a specialist in spectroscopy, he looked for it in experiments. However, the two Copenhageners found nothing and then shelved the idea. All that came out of their efforts was an interesting but unpublished draft manuscript.

Bohr only mentioned the isotope effect in public at a meeting of the British Association for the Advancement of Science in September 1915. On this occasion, he stated that the predicted shift for the neon isotopes Ne-22 and Ne-20 was $\Delta\lambda = 2.5 \times 10^{-6}$ cm and possibly detectable. However, neither this nor other isotope shifts were detected until much later. Stimulated by Bohr's idea, in the early 1920s spectroscopists looked for the effect in Li-6 and Li-7, but also in this simple case the expected atomic isotope shift failed to turn up.

On the other hand, the isotope shift in molecules was discovered in 1920 by physicists who succeeded in separating the vibrational frequencies in HCl due to the isotopes Cl-35 and Cl-37 which had recently been detected by Francis Aston by means of his new mass spectrometer. This kind of shift had been anticipated by Bohr five years earlier, before the chlorine isotopes were known. According to Bohr, the vibration frequency of a diatomic molecule consisting of masses M_1 and M_2 (as in HCl) would be proportional to the quantity $\sqrt{(M_1+M_2)/M_1M_2}$. The discovery of 1920 was important, not least because it led to Robert Mulliken's discovery five years later of the controversial zero-point energy in molecular vibrations (see also Section 4.4.2). However, the work in molecular spectroscopy did not involve quantum transitions between stationary states and therefore did not count as a confirmation of Bohr's original idea.

Strangely, the possibility of H-3 was discussed before the possibility of H-2. The standard view through the 1920s was that hydrogen has no heavier isotope or at least none abundant enough to turn up in experiments. Only in 1931 was the existence of H-2 seriously considered, and at Columbia University Harold Urey decided to look for it using the atomic isotope shift method. Urey and his collaborators, George Murphy and Ferdinand Brickwedde, made careful measurements of four of the hydrogen lines in the Balmer series using commercial hydrogen as well as samples enriched with the heavy isotope by evaporating liquid hydrogen. In both cases, they obtained distinct if faint isotope lines, and as expected less faint with the enriched samples. Importantly, the measurements agreed perfectly with the theoretical values. Thus, for the H_α line ($n = 3$ to $n = 2$) they found $\Delta\lambda = 1.79$ Å as compared to the predicted $\Delta\lambda = 1.787$ Å. As regards the relative abundance of H-2, they estimated it to be 1:4000 in natural hydrogen and 1:1000 in the enriched samples.

In 1934, Urey was awarded a full Nobel Prize in chemistry for his discovery of heavy hydrogen or, as it came to be known, deuterium. The work that led to the discovery was indirectly guided by Bohr's old theory, although at the time it had been replaced by the new atomic theory based on quantum mechanics. The latter theory gives the same expression for the isotope shift as the older one. In 1923, Urey had studied for a year at Bohr's institute in Copenhagen, but he most likely was unaware of Bohr's early and half-forgotten prediction of the isotope effect, to which he and his collaborators did not refer in their discovery paper. On the other hand, in his Nobel lecture, Urey mentioned that "Bohr's theory, given some twenty years ago, permits the calculation of the Balmer spectrum of the heavier isotopes of hydrogen" (Urey, 1934).

As Urey's reference to "heavier isotopes" (and not "heavier isotope") suggests, the existence of H-3 or tritium in addition to H-2 or deuterium was still considered a possibility. As it turned out, tritium does not exist naturally except as trace amounts in the atmosphere due to cosmic rays. The isotope is beta radioactive with a half-life of 12.3 years, and it was first isolated in 1939 by bombarding deuterium with high-speed deuterons from a cyclotron. The reported process was

$$^{2}_{1}\mathrm{H} + ^{2}_{1}\mathrm{H} \rightarrow ^{4}_{2}\mathrm{He} \rightarrow ^{3}_{1}\mathrm{H} + ^{1}_{1}\mathrm{H}.$$

Bohr was undoubtedly the first to hypothesize the tritium isotope, although he did not publish his suggestion. From the 1930s onwards, the isotope effect and related methods of separating and studying isotopes developed into a wide-ranging research area of great importance not only to physics and chemistry but also to astronomy, biology and the geological sciences.

1.2.5. MOLECULES AND MANY-ELECTRON ATOMS

In a letter to Hevesy during early February 1913, Bohr told him about his still immature ideas of a new atomic theory. "I have hope to obtain a knowledge of the structure of the systems of electrons surrounding the nuclei in atoms and molecules," he wrote, "and thereby hope of a detailed understanding of what we may call the 'chemical and physical' properties of matter" (Kragh, 2012, p. 74). He further promised that his theory would include "a very suggestive indication of an understanding of the periodic system of the elements."

Considerations and results of a chemical nature occupied an important if often unappreciated place in Bohr's trilogy, such as indicated by its title with "molecules" enjoying the same status as "atoms." Whereas Bohr focused on the hydrogen atom and the general principles in the first part, he devoted the second part mostly to many-electron atoms. In the third part, he ambitiously suggested models of simple molecules such as H_2, O_2, O_3, HCl, H_2O, and CH_4. Bohr's concern with molecules and elements mainly related to two areas, both of which were central to chemical thinking. One was molecular structure and the chemical bond, and the other was the periodic system of the elements. While his work in the first area was short-lived and largely unsuccessful, his work in the latter constituted for a period of time an important advance in theoretical chemistry.

According to Bohr, the nonpolar or covalent bond keeping two atoms together in a molecule could be modelled as a ring of two or more revolving electrons common to the two atoms. For example, his model of methane, CH_4, was tetrahedral, with the four C–H bonds represented by four rings each of which carried two electrons. He placed the remaining two electrons in a small circular orbit around the carbon nucleus. The model had an appealing affinity with what chemists knew about carbon compounds, but only because it was designed to mirror chemical knowledge. It was a qualitative model, not based on calculations of any sort.

In the case of the simple H_2 molecule — the only one Bohr analyzed quantitatively — the two electrons circled on a ring placed symmetrically between the nuclei. A straightforward calculation based on his theory of the hydrogen atom proved that the process $H + H \rightarrow H_2$ was exothermic with a heat of formation equal to 63 kcal/mole. Given that the American chemist Irving Langmuir had found experimentally a value of 76 kcal/mole, Bohr was understandably optimistic that he was on the right track. Alas, the optimism did not last, for Langmuir's improved measurements of late 1913 resulted in 84 kcal/mole with only a small error margin. There was no realistic hope of reconciling the experimental and theoretical values. After this disappointment, Bohr realized that his model of the covalent bond was inadequate and he chose to shelve it. Other physicists in the tradition of the old quantum theory continued to study the H_2 molecule, but their elaborate calculations failed to agree with experiments. The same was the case with the even simpler H_2^+ ion consisting of just two protons and one electron.

For atoms with many electrons, Bohr concluded in 1913 that the electrons revolved in a system of concentric orbits or "rings" and that the number of electrons in the outermost ring determined the element's valence and its chemical properties generally. Thus, for nuclear charges $Z = 3$, 11, and 19 he wrote the electron configurations as (2, 1), (8, 2, 1), and (8, 8, 2, 1), respectively, an indication that the corresponding elements Li, Na, and K belonged to the first group of the periodic table. The connection between chemical properties and the outermost electrons may seem obvious, but at the time it was not, and it first appeared in Bohr's theory. Also, Thomson, in his earlier atomic model prior to the discovery of the atomic nucleus, had suggested an explanation of the periodic system in terms of electron configurations. But in Thomson's model, the similarity between elements was the result of common *internal* groups of electrons.

The way in which Bohr assigned tentative electron configurations was, to some extent, based on considerations of mechanical stability and use of principles of quantum theory. However, empirical reasoning based on the known chemical and physical properties of the elements was far more important. In cases where the two approaches — one deductive and the other inductive — led to different results, he usually gave priority to the chemical-inductive approach. This eclectic and somewhat opportunistic method turned up, for example, in his treatment of the lithium atom with $Z = 3$. For the two-ring system (2, 1) Bohr calculated the total binding energy to −218 eV, whereas he got −240 eV for the (3) model with three electrons in a single ring. The latter model would thus be energetically favored, but he nonetheless concluded that the structure of the lithium atom was (2, 1) and not (3). The reason was of course that the first configuration agreed with lithium's chemical properties whereas the latter was incompatible with them. Bohr frankly admitted that in general "the number of electrons in this [outermost] ring is arbitrarily put equal to the normal valency of the corresponding element" (Bohr, 1913).

In the second part of his trilogy, Bohr applied his eclectic method to propose electron configurations from $Z = 1$ to $Z = 25$ that had a suggestive similarity to the periodic system. He cautiously emphasized that the configurations were tentative and only the beginning of what hopefully would be developed into a more complete explanation of the system. This more complete theory he presented in a series of works between 1920 and 1923, at a time when he had greatly improved the original atomic theory. Now the electrons moved in three-dimensional elliptical orbits with an eccentricity given by the ratio n/k, where k is the azimuthal quantum number with values $k = 1, 2, \ldots, n$ (corresponding to what in quantum mechanics is usually denoted l, with values from 0 to $n - 1$).

The system of electron orbits was much more complicated in the revised model, among other reasons because electrons moving in outer orbits might sometimes penetrate the inner core of the atom. Based on this more sophisticated model, and relying on data from X-ray spectroscopy and arguments derived from the correspondence principle, Bohr offered an atomic explanation of the entire system from hydrogen to uranium. Although his arguments were far from transparent, he was confident that now he had understood the fundamental principles of atomic architecture and at the same time shown how chemistry could be understood in terms of atomic and quantum theory. His confidence was premature.

Without going into the details of Bohr's theory of the periodic system, two of his results merit brief attention. First, by 1922 the nature and existence of element 72 was controversial. French scientists believed to have found traces of the element in the form of "celtium" in rare earth metals, but Bohr argued theoretically that the element must be a homolog of zirconium with atomic number 40. In $Z = 72$, there would be two outer electrons with $(n, k) = (5, 3)$ and two with $(6, 1)$, similar to zirconium's outer electrons in quantum states $(4, 3)$ and $(5, 1)$. The prediction was dramatically confirmed in early 1923, when element 72 was discovered in Copenhagen by means of X-ray spectroscopy and then named hafnium. For a while the celtium claim was maintained, but after a year or so it was recognized to be erroneous (see also Section 2.4.2).

Second, although Bohr did not believe that transuranic elements existed or could be manufactured, as a scientific exercise he extended his electron configurations to these still hypothetical elements. In his Nobel lecture of 1922, he explained that the imaginary element 118 would be an inert gas with electrons arranged as (2, 8, 18, 32, 18, 8). Nearly a century later, the element ceased being hypothetical. In 2016, the International Union of Pure and Applied Chemistry (IUPAC) recognized that a few nuclei of element 118 had been produced in nuclear reactions; hence, it was a real element and was assigned the name oganesson and chemical symbol Og. For more on element 118, see Section 2.5.5. Calculations indicate that the electron configuration of oganesson is (2, 8, 18, 32, 18, 8), the very same as Bohr suggested almost a century ago! The agreement or coincidence is remarkable also from a philosophical perspective. Because, how can a wrong theory based on orbiting electrons, and with no spin and no Schrödinger equation, give the very same result as modern quantum-chemical computations?

With few exceptions, chemists did not welcome Bohr's bold attempts to extend his theory to the domain of chemistry. They found the Bohr model to be utterly unconvincing from a chemical point of view, in particular because of its inability to account for valence and the structure of molecules. The American physical chemist Richard Tolman — who may be better known for his important contributions to theoretical cosmology (Chapter 4.3) — gave voice to the chemists' dissatisfaction. In an address of 1922, he spoke of the chemists' "extreme hostility to the physicists, with their absurd atoms, like a pan-cake of rotating electrons, an attitude which is only slightly modified by a pious wish that somehow the vitamin 'h' ought to find its way into the vital organs of their own, entirely satisfactory, cubical atom" (Tolman, 1922). What Tolman called the chemists' hostility to the physicists was not a new phenomenon, such as we shall see in Chapter 2.4.

The cubical atom mentioned by Tolman was favored by a majority of chemists. According to this model as proposed by Lewis and further developed by Langmuir, the electrons occupied fixed positions in the corners of an atomic cube or several concentric cubes. The covalent bond was pictured as a pair of electrons that shared an edge or a face in the adjacent atomic cubes. This kind of model was chemically useful but inadmissible from the point of view of Bohr and other quantum physicists, according to whom the atom must necessarily be a dynamical system. There was in the period from about 1918 to 1925 an extended controversy between two rival conceptions of atomic and molecular structure, one being dynamical-physical and the other static-chemical. None of the conceptions were quite satisfactory, and attempts to merge them did not succeed. Only with the celebrated work of Walter Heitler and Fritz London in 1927 was the covalent bond in the hydrogen molecule explained in terms of quantum mechanics, and with the later works of Linus Pauling and others, a kind of synthesis of the views of the chemists and the physicists was finally achieved. Whereas Bohr's model did not survive the transition to quantum mechanics, in a sense Lewis' did.

As far as the periodic system is concerned, Bohr's theory was a big step forward, but still only a step on the bumpy road toward a full understanding. The first modern quantum explanation of the periodic system was offered by Wolfgang Pauli in a classical paper of March 1925 in which the famous exclusion principle was first enunciated. Contrary to Bohr's pictorial theory based on electron orbits, Pauli's theory was of a purely formal nature and made use of four rather than two quantum numbers. One of them was the spin quantum number, but only if seen in retrospect. After all, the discovery of the electron's spin was announced in October 1925, much later than Pauli's theory and shortly after the emergence of Werner Heisenberg's new quantum mechanics. It took a while until Pauli accepted the electron's spin and realized that it expressed the same classically non-describable *Zweideutigkeit* (ambivalence, but here double meaning) which he had introduced as a formal principle half a year earlier.

.1.3

From Positrons to Antimatter

1.3

From Positrons to Antimatter

The material world consists of matter made up of electrons, protons, and neutrons. The latter two particles are also known as nucleons — nuclear particles — and they are built up of elementary quarks. According to fundamental theory, each of these well-known particles, as well as other elementary particles, must exist in the symmetric form of antiparticles such as positively charged electrons (or positrons) and negatively charged protons (or antiprotons). Antiparticles are usually denoted by placing a bar above the symbol, thus $\bar{p}$ for the antiproton and $\bar{n}$ for the antineutron; the exception to the rule is the positron, where the symbol is e^{+} rather than $\bar{e}$. The antineutron is electrically neutral just like the ordinary neutron n, but the two particles differ in other respects. These antiparticles have long been known to exist, but they are exceedingly rare in nature, and antiatoms made up of antiparticles are even more rare if existing at all. Given the fundamental symmetry of particles and antiparticles, this poses a deep and, still to this day, poorly understood mystery, a problem where particle physics and cosmology share a common ground.

By far, the best known antiparticle is the positron, the only one which has been granted its own name. According to the standard summary history as presented in textbooks and encyclopedia, the positive electron is just another name for the positron, a particle predicted by the young quantum wizard Paul Dirac in 1931.

The following year, Dirac's theoretical antielectron or positron was discovered in cosmic-ray experiments and hence turned into a real particle. However, this standard history is flawed in several respects and disregards, for example, much earlier discussions of positive electrons. Moreover, from a modern point of view the identity of positive electrons and antielectrons is taken for granted, and the same is the case with negative protons and antiprotons. Physicists in the 1930s were not so sure. For a period of time, they recognized the existence of positrons without recognizing the existence of antielectrons. As usual, authentic history presents a more complex and messy picture of the development than does the stream-lined and sometimes anachronistic textbook history.

The term "antimatter" first became commonly used in the 1970s when physicists began to wonder about atoms made up of antiparticles. Curiously, "antimatter" appears as early as 1898, when the German-born British physicist Arthur Schuster coined the name in a whimsical paper published in *Nature*. Schuster speculated that there might exist somewhere in the universe a hitherto unknown form of matter with the remarkable property that it would be repelled gravitationally by ordinary matter. "Worlds may have formed of this stuff," he wrote, "with elements and compounds possessing identical properties with our own, undistinguishable in fact from them until they are brought into each other's vicinity" (Schuster, 1898). Schuster's hypothetical antimatter, relating to antigravity and not to electrical charges, has only the name in common with the later concept of antimatter. Besides, he was well aware that his "holiday dream" (as he called it) was nothing but an innocent speculation.

1.3.1. POSITIVE ELECTRONS BEFORE THE ANTIELECTRON

The modern electron has two fathers, both of them Nobel laureates. In 1896, the Dutch physicist Pieter Zeeman discovered the effect eponymously named after him, namely that a spectral line is split in two or more lines when the source of light is placed in a magnetic field. Inspired by calculations of his compatriot H. A. Lorentz, he inferred from the observed splitting that matter contains vibrating electrons with a charge-to-mass ratio e/m approximately 1000 times larger than that of the hydrogen ion H^+ known from electrolysis. The following year J. J. Thomson demonstrated in a celebrated series of experiments that the enigmatic cathode rays consist of negatively charged subatomic particles with the same e/m ratio. Thomson boldly hypothesized that the electron (which he called a "corpuscle") was the basic constituent of all matter. Whether in Zeeman's or Thomson's version, and irrespective of the name they used for the particle, the electron turned up only with a negative charge.

By that time, not only were electrical subatomic particles well known theoretically, so was the name "electron," which in 1891 was introduced by the Irish physicist George Johnstone Stoney as an electrical unit charge instead of the "electrine" he had suggested in 1874. Stoney's electrons were parts of the atom, but they were not assigned any specific mass or sign of charge, nor could they be removed from the atom. The idea that the atoms of matter are conglomerates of positive and negative charges goes back to the 1830s and was cultivated in particular by German physicists. According to Wilhelm Weber, a reputed expert in electrodynamics, matter as well as ether consisted of hypothetical electrical particles with equal but opposite charges. Positive and negative electrons also appeared in the theories of the distinguished British mathematical physicist Joseph Larmor, who referred to "electrons" as singularities in the electromagnetic ether. In an important treatise of 1894, Larmor stressed that the only difference between these two primordial particles was the sign of the charge. Atoms, he wrote, "may quite well be composed of a single positive or right-handed electron and a single negative or left-handed one revolving round each other" (Larmor, 1894).

Although positive electrons as mirror particles of the negative electrons obstinately failed to turn up in experiments, still around 1900 they were thought of as respectable and possibly real constituents of matter and ether. In a paper of 1901, the Australian-British physicist William Sutherland proposed a picture not unlike the earlier ones of Weber and Larmor. He spoke of an elementary particle composed of a positive and negative electron revolving around their center of gravity. It is almost as if Sutherland and his predecessors speculatively anticipated what much later was called positronium (see Section 1.3.3). A more elaborate atomic model somewhat similar to Sutherland's was proposed by 24-year-old James Jeans, who explicitly mentioned that the two kinds of electrons were mirror particles. Moreover, he speculated that two oppositely charged electrons might "rush together and annihilate one another" (Jeans, 1901). Here we have another possible anticipation. Twenty years later, another young British physicist, Paul Dirac, predicted the e^+e^- annihilation process on the basis of his novel and completely non-classical idea of antielectrons.

At about the time of the speculations of Sutherland and Jeans, the positive electron lost credibility and was soon abandoned. By 1905, it had become an unwelcome particle, not only because of the lack of evidence for it but also because its existence disagreed with the successful research program in electron physics building on negative electrons only. The standard view was summarized by the Cambridge physicist Normann Campbell, who in a book of 1907 stated, "if there is one thing which recent research in electricity has established, it is the fundamental difference between positive and negative electricity" (Campbell, 1907, p. 130). Although the positive electron was abandoned, somewhat confusingly, the name continued to be popular, if now in a different meaning. The positive electrons appearing frequently in the physics literature ca. 1905–1925 were no longer mirror particles of the electron. They were hydrogen ions H^+ or what after 1920, when Ernest Rutherford introduced the name, were increasingly called protons. As late as 1935, after the discovery of the positron, Robert Millikan wrote a book with the title *Electrons (+ and –), Protons, Photons, Neutrons, and Cosmic Rays*. Although the name "proton" thus appeared in the book, it was limited to the title. Throughout the text, Millikan consistently referred to protons as positive electrons. Most likely, many readers were confused.

It was known that the heavy proton (or so-called positive electron) had exactly the same numerical charge as the light negative electron, but few mainstream physicists considered it a puzzle worth contemplating or in need of explanation. Young Wolfgang Pauli was one of the few. In a letter to Arthur Eddington of 1923, he expressed his dissatisfaction with current atomic and electron theory: "Why are there only two kinds of elementary particles, negative and positive electrons (hydrogen nuclei)? Why have their electrical charges the same value, if their masses are so very different?" (Kragh, 1989a). Neither Pauli, Eddington, nor other theoretical physicists could come up with an answer. In a paper in *Nature* in 1922, Oliver Lodge suggested that the proton might be built up of a large number of $n + 1$ positive electrons and n negative electrons, but at the time his speculation was not taken seriously by mainstream physicists. Still, in the light of Dirac's hypothesis of 1929 to be discussed below, Lodge's suggestion is of some interest. His aim was the same as Dirac's, namely, to build up all matter of electrons alone.

Not all physicists agreed with the consensus view that the true positive electron, carrying a mass equal to that of the negative one, could be ruled out. For a brief period of time, experimental evidence seemed to indicate the existence of positive electrons, such as argued in particular by the French physicist Jean Becquerel, a son of Henri Becquerel of radioactivity fame. Becquerel Junior based his discovery claim on a series of experiments conducted between 1907 and 1910 in which he believed to have found convincing evidence of both bound and free positive electrons. In one line of research, he observed an "opposite Zeeman effect" corresponding to vibrations of positive rather than negative unit charges. He believed that the new effect was strong evidence for positive electrons in matter, and his confidence grew when he also found the free particles in experiments with discharge tubes. However, other experts were unable to repeat his experiments, which they denied provided evidence for positive electrons. Although Becquerel's discovery claim received support from a few physicists, among them the later Nobel laureate Heike Kamerlingh Onnes, it was generally rejected.

Without conceding defeat, after 1910, Becquerel silently withdrew from the controversy, which was soon forgotten. After all, physicists knew, or thought they knew, that matter consisted of only negative electrons and positive protons and that the hypothetical positive electron was not part of the fabric of nature. Apart from a few speculations of a philosophical rather than scientific nature, such as Lodge's, the hypothesis of a positive counterpart to the electron was absent from physics in the 1920s. Curiously, in 1923 the American engineer and amateur philosopher Arvid Reuterdahl suggested the name "positon" (without an r) for what he considered the negative electron's mirror particle.

The suggestion made no impact at all on the physicists, most of whom were unaware of Reuterdahl and his ideas. When the positive electron returned to physics in 1931, its mass and charge were the same as discussed in the first decade of the century and yet it was a very different particle. Initially, Dirac's hypothetical antielectron was as unwelcome as its classical predecessor, but contrary to this particle it turned out to be real.

1.3.2. A MIRACULOUS EQUATION

By the summer of 1927, quantum mechanics was essentially complete. In its standard formulation, it was based on the famous wave equation Erwin Schrödinger announced in the spring of 1926 and which describes the behavior of an electron in terms of a wave function ψ depending on space and time. The Schrödinger equation can be considered as a quantum translation of the classical energy equation, which for a free particle with momentum p moving in one dimension is

$$E = p_x^2/2m.$$

The recipe for the quantum translation is given by the substitutions

$$E \to i\hbar\frac{d}{dt} \quad \text{and} \quad p_x \to -i\hbar\frac{d}{dx},$$

where the differential operators act on the wave function. The symbol $\hbar$ is an abbreviation for $h/2\pi$ and i denotes the imaginary unit with the property $i^2 = -1$. As Schrödinger was acutely aware, his equation was non-relativistic and thus neither generally applicable nor truly fundamental. According to the special theory of relativity, the energy–momentum equation in one dimension reads

$$E^2 = p_x^2c^2 + m^2c^4,$$

where c is the speed of light and m now denotes the rest mass of the particle. Relativity theory requires momentum and energy to be of the same order, to appear as E^2 and p^2, which in quantum mechanics suggests the quantities $d^2\psi/dt^2$ and $d^2\psi/dx^2$. In fact, Schrödinger originally derived a relativistic equation of this kind, but when he realized that it disagreed with the details of the hydrogen spectrum, he fell back on the non-relativistic approximation. Schrödinger's original equation was first published by other physicists and is known as the Klein–Gordon equation in honor of Oskar Klein and Walter Gordon, a Swedish and a German physicist.

Unfortunately, the theoretically appealing Klein–Gordon equation was unappealing when it came to explaining the physics of electrons and atoms. At the time there was another problem, which many physicists saw as connected to the relativity problem, albeit in a way they could not figure out. Since the fall of 1926, it had been known that the electron possesses an intrinsic angular momentum, a spin quantum number of value either ½ or –½. In order to incorporate the spin in the formalism of quantum mechanics, Pauli suggested to represent the electron's spin states by 2×2 matrices with two rows and two columns. It followed that the wave function must be doubled, written as (ψ_1, ψ_2) and not just as a single ψ as in Schrödinger's theory. By adding to the Schrödinger equation two terms, one relating to the spin and the other to a correction from relativity theory, Pauli and other physicists constructed an equation that accounted for most experimental data. However, this empirically satisfying equation was still of the Schrödinger type with energy and momentum appearing in different orders and therefore not in accordance with the theory of relativity.

The British physicist Paul Adrian Maurice Dirac was one of the founders of quantum mechanics, which he, at the age of 23, had established in his own version in the fall of 1925. Dirac wanted to find a fundamental quantum equation that was fully relativistic and also satisfied the formal structure of quantum mechanics, which to him meant that it must involve d/dt as the operator for energy. This ruled out the Klein–Gordon equation and its d^2/dt^2 operator. When Dirac addressed the problem at the end of 1927, he was of course aware of the spin problem, but for reasons of simplicity he decided to ignore it and consider at first only an electron without spin, obviously a toy model.

Dirac's reasoning led him to consider the expression

$$E/c = \sqrt{p_1^2 + p_2^2 + p_3^2 + (mc)^2},$$

where energy and momentum are represented by the corresponding differential operators. Since E is of the first order in the time derivative, the right side of the equation must be a linear combination of the quantities under the square root. In other words, it must be of the form

$$\alpha_1 p_1 + \alpha_2 p_2 + \alpha_3 p_3 + \alpha_4 mc.$$

Do there exist α-coefficients of such a kind that they turn the square root of a sum of four squares into a linear form? Guided by Pauli's spin matrices and his own mathematical intuition, Dirac realized that the coefficients had to be 4×4 matrices. As the Pauli matrices necessitated a wave function with two components, the Dirac matrices necessitated a four-component wave function:

$$\psi = (\psi_1, \psi_2, \psi_3, \psi_4).$$

With this insight, Dirac formulated his new equation, unifying quantum mechanics and special relativity, not only for a free electron but also for an electron placed in an electromagnetic field. It turned out, much surprisingly, that without introducing the spin beforehand, the correct spin of the electron came out of the equation, which of course was a great triumph. Moreover, Dirac's theory agreed perfectly with the hydrogen spectrum and other spectroscopic data.

Dirac's linear quantum equation was hailed as a great advance, but its physical interpretation immediately raised problems. First of all, whereas two of the wave functions, say ψ_1 and ψ_2, refer to the spin states of the electron, what do ψ_3 and ψ_4 refer to? In his paper of 1928, Dirac pointed out that in a formal sense the two extra functions were associated with the nonsensical idea of negative-energy electrons or, alternatively and not quite as nonsensical, electrons carrying a positive charge. Both interpretations were unacceptable, if for different reasons. The energy of a real object is necessarily positive, and as regards the electrical charge it was known that electrons are always negative. Dirac was thus faced with a dilemma of choosing between something absurd (negative energy) and something non-existing (positive electrons).

In the late autumn of 1929, Dirac thought that he had found an answer to the dilemma, which he first communicated to Bohr in a letter of 24 November. Imagine, he wrote, that there exists a bottomless sea uniformly filled with an infinite number of electrons with energies ranging from $-\infty$ to $-mc^2$. In that case the negative-energy electrons will be unobservable and merely define a vacuum state. Dirac further imagined that a few of the negative-energy states might be unoccupied, constituting a hole or vacancy in the sea of negative energy. "Such a hole would appear experimentally as a thing with +ve energy," he told Bohr (Kragh, 1990, p. 91). "These holes I believe to be the protons. When an electron of +ve energy drops into a hole and fills it up, we have an electron and proton disappearing simultaneously and emitting their energy in the form of radiation." That is, rather than identifying the extra variables in his equation with a positive electron, he chose to identify them with the proton, the only known elementary particle with a positive charge. The predicted annihilation process was the hypothetical

$$p^+ + e^- \rightarrow \text{radiation.}$$

According to modern physics, the process is forbidden, but this was not known at the time, when it was considered a possibility in stellar energy reactions (Section 3.3.3). Annihilation of matter, possibly of the proton-electron type, also featured in a cosmological model proposed by Richard Tolman in 1930.

Given that the proton is nearly 2000 times heavier than the electron, Dirac's hypothesis was problematic, but it appealed greatly to him for philosophical reasons. When he presented the theory in public, he emphasized how attractive it was to explain all matter in terms of a single elementary particle:

> "It has always been the dream of philosophers to have all matter built up from fundamental kind of particle, so it is not altogether

> satisfactory to have two in our theory, the electron and the proton. There are, however, reasons for believing that the electron and proton are really not independent, but are just two manifestations of one elementary kind of particle" (Dirac, 1930).

A statement to the same effect also appeared at the end of Dirac's influential textbook *Principles of Quantum Mechanics*, the first edition of which appeared in the summer of 1930. The book was highly successful and soon translated into even more successful German and Russian editions. However, Dirac's general view of quantum mechanics was dangerously close to what would be later called the Copenhagen interpretation, and this raised the eyebrows of the Soviet commissars responsible for the publication. The Russian edition consequently warned in a foreword that "this work contains many views and statements completely at variance with dialectical materialism" (Kragh, 1990, p. 78). The commissars had rather seen quantum mechanics used for "a smashing attack on the theoretical front against idealism," something the abstract and mathematically oriented *Principles* did not offer.

Dirac maintained the unitary hypothesis for more than a year, and that in spite of its obvious problems. Not only was there no evidence for electron–proton annihilation, it was also impossible to explain why the proton is so much heavier than the electron. When Dirac conceded his failure and decided to replace the real proton with a hypothetical positive electron, he kept to his imagery of holes in a sea of negative-energy electrons. In a seminal paper of June 1931, he briefly introduced the idea of mirror- or antiparticles in the form of a new elementary particle, a positive electron: "A hole, if there were one, would be a new kind of particle, unknown to experimental physics, having the same mass and opposite charge to an electron. We may call such a particle an anti-electron" (Dirac, 1931).

In his paper of 1931, Dirac introduced no less than three new hypothetical subatomic particles in addition to the two already known. One was the antielectron and another the antiproton (Section 1.3.4), but most of the paper was devoted to arguments justifying the existence of magnetic monopoles, magnetic analogues of electrons. Ambitiously, Dirac sought for an explanation of the elementary electrical charge e such as given by the dimensionless fine-structure constant

$$\frac{e^2}{\hbar c} \cong \frac{1}{137}.$$

Although Dirac did not succeed to derive the fine-structure constant theoretically, he proved that magnetic monopoles were allowed by quantum mechanics, which for him was a strong argument for their real existence (see also Section 4.2.4). Alas, contrary to the antielectron, the magnetic monopole has remained hypothetical to this day.

Dirac made it clear that he considered the positively charged antielectron a real particle and not just a mathematical abstraction. He suggested that just as an electron and an antielectron might annihilate ($e^- + e^+ \rightarrow$ radiation), so might the opposite process take place, the formation of an electron–antielectron pair out of radiation energy: radiation $\rightarrow e^- + e^+$. Since there was no law of nature forbidding the two processes, Dirac concluded that they would probably "occur somewhere in the world." However, the prediction of antielectrons or positive electrons was met with incredulity by most of Dirac's colleagues. The general attitude was that the daring hypothesis was interesting but too speculative to be taken seriously.

1.3.3. ANTIELECTRON OR POSITRON?

Many physicists considered Dirac's prediction of a positive electron to be as poorly justified as his earlier proton hypothesis. Both ideas were based on the hole theory, which was widely regarded with skepticism or as merely a provisional picture. In Copenhagen, the negative-energy electrons were jocularly known as "donkey electrons," a name suggested by George Gamow, because electrons with negative energy, like donkeys, would move slower the harder they were pushed! The donkey electrons turned up in a memorable parody of Goethe's *Faust* staged at Bohr's institute in 1932. The occasion was not only the tenth anniversary of the institute but also the centennial of Goethe's death. Bohr played the role of God (of course he did), Pauli was Mephistopheles, and among the other participants were Dirac, Gamow, Heisenberg, and Klein. At a time when the positive electron had not yet been discovered, Dirac presented his hole theory as follows:

"That donkey-electrons should wander
Quite aimless through space, is a slander,
That only with articles
On hole-like particles
Could be said to have found a defender" (Beller, 1999).

The skeptical Bohr replied with a reference to the potential annihilation catastrophe, the end of the world if antielectrons were real and as abundant as electrons:

"But the point of the fact is remaining
That we cannot refrain from complaining,
That such a caprice
Will reveal the malice
Of devouring the world it's sustaining."

The negative attitude to Dirac's theory was strongest before the positive electron was discovered, but even after the discovery, the theory was severely criticized by Bohr, Pauli and several other physicists. After all, was the positive electron the same as Dirac's antielectron? In 1933, Pauli published an authoritative review of quantum mechanics titled *Die Allgemeinen Prinzipien der Wellenmechanik* (General Principles of Wave Mechanics) in which he rejected the idea of strictly symmetric electrons and antielectrons as unworthy of serious consideration.

At the time that Dirac proposed his interpretation of the negative-energy states appearing in his theory, physicists in California were busy with experimental studies of the enigmatic cosmic rays hitting the Earth from outer space. Robert Millikan and his coworkers believed that the primary cosmic rays were high-energy photons and that the observed charged particles in the upper atmosphere were caused by atomic nuclei when hit by one of those photons. This was a view shared by Millikan's former student Carl Anderson. Using cloud chambers sent up with balloons, in 1932 Anderson found tracks in the cloud chamber photographs which he identified as due to positive particles in the cosmic rays.

According to physics textbooks, encyclopedias, and website articles, not only did Anderson in this way discover the positive electron, he was also guided by and confirmed Dirac's prediction of antielectrons.

Four years later, Anderson was awarded the Nobel Prize, and on this occasion the spokesman for the Nobel Prize committee, the Swedish physicist Henning Pleijel, added support to what is largely a myth:

> "Dirac formulated the hypothesis that it might be that in other parts of the universe positive and negative charge were reversed. Dr. Anderson now pursued his investigations ... and after having carried out verifying experiments and new measurements he was able to furnish, in the summer of 1932, clear evidence of the existence of the positive electron. The positron Dirac had been searching for was thus found."[1]

The authentic story is quite different from the one usually presented, and it is much less morally suited for textbooks. For one thing, Anderson at first interpreted the tracks as due to protons and only later concluded that the particles carried a mass comparable to that of the electron. This is what he reported in a paper in *Physical Review* of March 1933, where he announced the discovery of what he called the "positron." (He also suggested the logically reasonable "negatron" for the ordinary electron, but the name was never widely used.) Moreover, Anderson's understanding of the positron was based on classical physics and not on quantum mechanics, and it was not indebted at all to Dirac's theory to which he did not refer and was only vaguely aware of. In agreement with Millikan's old-fashioned ideas, he thought of protons and electrons as tiny electromagnetic spheres with mass and radius given by

$$m = \frac{e^2}{c^2 r}.$$

As Anderson wrote in the discovery paper, "an encounter between the incoming primary ray [a high-energy photon] and a

[1] https://www.nobelprize.org/prizes/physics/1936/ceremony-speech/.

proton may take place in such a way as to expand the diameter of the proton to the same value as that possessed by the negatron" (Anderson, 1933). In brief, Anderson discovered the positron, but not the antielectron, and his discovery of the positron was serendipitous, not planned. The identification of the positron as an antielectron was made clear in roughly contemporary cloud chamber experiments by Patrick Blackett and Giuseppe Occhialini, who, contrary to Anderson, were acquainted with Dirac's hole theory. The two Cambridge physicists demonstrated for the first time the pair creation predicted by Dirac. With this work, the antielectron gained in credibility and was soon accepted as real, although many physicists preferred to think of the positron as separate from the antielectron.

With the discovery of artificial radioactivity in early 1934, it was realized that positrons are neither rare nor particularly exotic. They could be produced at will. With the new picture of the atomic nucleus — now a proton–neutron composite rather than a proton–electron composite — the phenomenon was interpreted as a nuclear proton transforming into a neutron. The decay was not only accompanied by a positron but also by another new elementary particle, the neutrino ν which Pauli had predicted in 1930 and originally conceived as a constituent of the atomic nucleus. The modern formulation of the process is

$$p \rightarrow n + e^{+} + \nu.$$

Whereas positrons became popular and even practically useful, few of the early texts on artificial radioactivity mentioned antielectrons. Positron-emitting radioactive isotopes have long found use in hospitals and biomedical laboratories. More recently, they have turned up in PET scanning technologies, PET being an acronym for positron-electron tomography. It is truly amazing that Dirac's old prediction based on his relativistic wave equation has morphed into a medical imaging instrument widely used in cancer treatment and other clinical areas.

Positron emission is not limited to artificially produced isotopes but also occurs in nature. Ordinary potassium with atomic weight 39.1 contains 0.01% of the heavy isotope K-40, such as George Hevesy showed in 1935. The isotope turned out to be radioactive with a lifetime of approximately 1.3×10^9 years, of the same order as the age of the Earth. It normally decays to Ca-40 by emitting a negative electron (β radioactivity), but as became established around 1945, at rare occasions it emits a positron and transmutes into Ar-40. Although this is a very rare process, it means that in an average human body a few thousand positrons are emitted every second. And it happens quite naturally, without we sensing it or being able to do anything about it.

While Dirac predicted electron–positron annihilation, he did not consider the possibility of an unstable atomic system made up of the two particles. In a paper of 1934, an obscure Yugoslavian physicist by the name Stjepan Mohorovičić was the first to speculate about such an exotic atom, which he associated with the no less exotic element "electrum," chemical symbol Ec. Mohorovičić pictured this lightest of all atoms (with atomic weight about 0.001) as an electron and a positron revolving in Bohr orbits around their common center of gravity. Moreover, he hypothesized more complex electron–positron atoms in the hope that they might be detected in stellar spectra. His paper was forgotten, and when the short-lived "positronium" was discovered in 1951, it was in ignorance of Mohorovičić's earlier prediction. Positronium in its ground state decays by annihilation to photons with a lifetime of the order 10^{-10} seconds, and it is only known from the laboratory. Nonetheless, this simple system is of great scientific interest to physicists and chemists.

1.3.4. THE NEGATIVE PROTON

From a modern point of view, the negative proton is just another name for the antiproton, a well-known particle produced routinely and copiously in high-energy laboratories. Symbolically, $p^- \equiv \bar{p}$. The situation in the 1930s was quite different. Although the hypothetical negative proton was occasionally discussed in the period, it was typically distinguished from the no less hypothetical and even less credible antiproton in Dirac's sense. In a letter to Heisenberg of 17 April 1934, Pauli wrote: "Bohr thought much about negative protons and believes to have found evidence for their existence in the cosmic rays. ... The relativistic Dirac wave equation is not at all applicable to heavy particles, and Bohr believes therefore that the negative protons *should not at all be related to the idea of holes and hence not annihilate with the positive protons!*" (Kragh, 1989b).

As far as the antiproton is concerned, the idea and name first appeared in Dirac's paper of 1931, in which he introduced the antielectron. Dirac suggested, if only briefly, that apart from the antielectron, the picture of unoccupied negative-energy states might be valid also for protons and "appearing as an anti-proton." Very few physicists accepted Dirac's suggestion, one of the few being the Russian–Italian physicist Gleb Wataghin, who in a paper of 1935 introduced for the first time the antineutron. As Wataghin pointed out, this particle would have the same zero charge as the neutron, but in agreement with Dirac's ideas it would annihilate with an ordinary neutron according to $\bar{n} + n \rightarrow$ radiation. Whereas the antiproton and antineutron were ignored by most physicists, negative non-Dirac protons, although no less hypothetical, appeared rather frequently in the period's scientific literature. If the negative proton existed as a real particle, it might be in the mysterious atomic nucleus or perhaps in the equally mysterious cosmic rays.

Gamow, a pioneer and authority in nuclear theory, believed for a while that negative non-Dirac protons resided in the nucleus. He thought that in this way he could explain the stability of the beryllium-9 nucleus, which at the time was considered a puzzle. Measurements showed that the mass of this nucleus exceeded the masses of two α particles and one neutron, which implied that the decay process

$$ {}^{9}_{4}\mathrm{Be} \rightarrow {}^{4}_{2}\mathrm{He} + {}^{4}_{2}\mathrm{He} + n $$

should occur spontaneously. The usual picture of the Be-9 nucleus was that it consisted of 4 protons and 5 neutrons, but Gamow, with the support of Pauli and Dirac, suggested that the nucleus might be composed of 5 protons, 3 neutrons, and 1 negative proton. In that case, there were not enough neutrons to form the two α particles, and the decay process would not occur. In about 1936, improved mass determinations made the puzzle disappear, as it turned out that the masses on the right side were in fact greater than the mass of Be-9. Hence there was no need to introduce a negative proton in the nucleus.

Although admitting that negative protons lacked experimental support, Gamow maintained his belief for several years. In *Structure of Atomic Nuclei and Nuclear Transformations*, a book of 1937, he summarized his arguments for negative protons, emphasizing that they were entirely different from Dirac's antiprotons. During the 1930s, negative protons (but not antiprotons) were occasionally discussed as constituents of cosmic rays. Millikan even thought to have found evidence for the particles. In late 1937, he reported to have found heavy electrical particles that might possibly be protons. "If they are protons, since they are positive and negative in sign, a negative proton has been discovered" (Kragh, 1989b). This is what Millikan said to *The New York Times*, but he soon got second thoughts and avoided making a formal discovery claim.

By the early 1950s, nuclear and particle physics had advanced greatly. Dirac's particle–antiparticle symmetry was recognized to be a fundamental principle of nature incorporated in a general invariance property of quantum fields called charge conjugation or *C*-invariance. Physicists were now convinced that negative protons were identical to antiprotons and that they existed somewhere in nature or at least could be produced artificially. In order to create the particles in collisions between accelerated beams of protons and matter, the energy of the protons had to be at least 5.6 GeV (1 GeV = 10^9 electron volts). The Berkeley accelerator known as the Bevatron was ready in 1954 and could accelerate protons up to 6.2 GeV, which made it the ideal machine.

In late 1955, a team of Berkeley physicists using the Bevatron found what they expected, namely antiprotons, although in the first experiments they were unable to observe the characteristic annihilation signature. The discovery paper started with the motivation of the experiment, namely, "until experimental proof of the existence of the antiproton was obtained, it might be questioned whether a proton is a Dirac particle in the same sense as the electron" (Chamberlain *et al.*, 1955). Now the question had been answered. Two years later, other Bevatron experiments revealed the antineutrino, and in 1959 the discovery of the antiproton resulted in a Nobel Prize to two of the Berkeley physicists, Emilio Segré and Owen Chamberlain. Although unsurprising in itself, the discoveries encouraged physicists to take antiparticles more seriously. It was quickly followed by astrophysical and cosmological ideas concerning antiparticles in space.

One thing is elementary antiparticles, another is antinuclei or antiatoms made up of separate antiparticles. A proton and a neutron stick together in the form of a deuteron d, the nucleus of the heavy hydrogen isotope. As two teams of physicists, one American and the other European, showed in 1965, the same is the case with the corresponding antiparticles. The discovery of the antideuteron $\bar{d}$ stimulated high-energy experiments aiming to produce and detect more complex composites of antiparticles. In 1995, physicists at CERN, the large European research facility near Geneva, succeeded in detecting the first antiatom ever, antihydrogen made up of a positron revolving around an antiproton. The handful of antiatoms was very short-lived, but with improved technology the CERN physicists were able to produce a large number of antihydrogen atoms and keep them alive for several minutes. When the breakthrough was announced in 2002, CERN was pleased to point out that it took place on the centenary of Dirac's birth.

One might imagine that it is just a matter of enough money and energy to build up heavier antiatoms, but it is not that easy. In 2011, physicists detected a bound system of two antiprotons and two antineutrons, an antihelium nucleus or antialpha particle. Attempts to add two positrons and thus form an antihelium atom have failed so far. Heavy antielements such as antigold and antiuranium belong to science fiction and not to science. So do antihumans.

1.3.5. COSMIC ANTIMATTER

When Dirac was awarded the Nobel Prize of 1933, sharing it with Schrödinger, it was for his contributions to quantum mechanics in general and not for his theory of antiparticles. Nonetheless, this was the theory he dealt with in his Nobel lecture titled "Theory of Electrons and Positrons." Without referring to Anderson's recent discovery of the positron, Dirac ended the lecture by extending the idea of antiparticles to a cosmological perspective:

> "We must regard it rather as an accident that the Earth (and presumably the whole solar system), contains a preponderance of negative electrons and positive protons. It is quite possible that for some of the stars it is the other way about, these stars being built up mainly of positrons and negative protons. In fact, there may be half the stars of each kind. The two kinds of stars would both show exactly the same spectra, and there would be no way of distinguishing them by present astronomical methods."[2]

It was only in the wake of the discovery of the antiproton that a few astronomers and physicists followed up on Dirac's cosmic speculation.

As pointed out by the Austrian–American nuclear physicist Maurice Goldhaber in a paper of 1956, all existing theories of physical cosmology made the arbitrary assumption that matter creation is not accompanied by the creation of antimatter. He may have been the first to raise the question of why there is so much matter in the universe and practically no antimatter. Given the fundamental symmetry between the two forms of matter, this was a mystery. In vain of better ideas, Goldhaber speculated that perhaps our cosmos has a counterpart in the form of an "anticosmos," which was separated off from our world at the beginning of time. More than sixty years later, the anticosmos returned to physics, if now separated from our universe in time and not in space. A cosmological model proposed in 2018 assumes a pre-big-bang antimatter universe symmetric to our post-big-bang matter universe. According to the three physicists behind this remarkable scenario, "The universe before the big bang and the universe after the big bang may be viewed as a universe/antiuniverse pair, created from nothing" (Boyle *et al.*, 2018).

At the time of Goldhaber's spirited speculation, nothing was known about the cosmic abundance of antimatter, except that it must be exceedingly small. The first rough estimate was obtained in 1962, when scientists used the artificial satellite Explorer XI to

---

[2] https://www.nobelprize.org/prizes/physics/1933/dirac/lecture/.

look for gamma photons of high energy arising from proton-antiproton annihilation. Only very few photons of this kind were found, which underlined the problem raised by Goldhaber. Cosmic antimatter turned up at an early date, also in connection with the steady-state theory, which in the 1950s was a serious rival to evolution theories of the big-bang type (Section 4.1.4). Steady-state theory assumed an eternally expanding universe, from which followed the hypothesis of continual creation of protons throughout space. But why protons and not an even mixture of protons and antiprotons? Fred Hoyle, the leading advocate of steady-state cosmology, considered the possibility of charge-symmetric matter creation as early as 1956. However, calculations indicated a background of annihilation gamma photons incompatible with observations, and Hoyle consequently shelved the hypothesis. Whereas the steady-state theory relied crucially on matter creation, it was of no importance whether the created matter involved antiparticles or not.

By the late 1960s, after the big-bang model of the universe had been generally accepted, the missing antimatter problem became a matter problem. In the very early universe, antinucleons were assumedly as abundant as nucleons, and almost all of these particles would annihilate into photons until the universe had expanded to such a size that annihilation became rare. The result would be a nearly complete annihilation of matter, an almost empty universe filled with photons. Obviously, this is very different from the universe we know, filled as it is with galaxies, cosmic dust, stars, and planets. Somehow the "annihilation catastrophe" had been prevented, but how? Some cosmologists ascribed the presence of matter to an initial asymmetry between matter and antimatter in the very early universe, but this was widely seen as explaining away the problem rather than explaining it.

Antimatter also played a crucial role in one of the alternatives to big-bang cosmology, which attracted attention in the 1970s. Developing earlier ideas of his compatriot Oskar Klein, the prominent Swedish physicist Hannes Alfvén argued in a series of papers that the universe contains equal amounts of matter and antimatter — mostly protons and antiprotons — separated by cosmic electromagnetic fields. As a result of gravitational condensation, matter–antimatter annihilation processes would increase and produce a huge radiation pressure that halted the contraction and eventually turned it into the observed expansion. Alfvén's so-called plasma cosmology assumed an eternal universe with no big bang and therefore did not need to explain the origin of protons and antiprotons. Although plasma cosmology is still defended by a few supporters, it has long ago ceased to be taken seriously by mainstream cosmologists sharing the big-bang paradigm.

Plasma cosmology helped spread the idea of antimatter known to a wider audience. In a popular book on his favorite cosmological model published in 1966, shortly after the discovery of the antideuteron, Alfvén expressed his firm belief in more complex forms of antimatter. Not only was he certain that atomic and molecular antihydrogen ($\bar{H}$, $\bar{H}_2$) existed, the same was the case with anti-water ($\bar{H}_2\bar{O}$) and more complicated chemistry. "Antihydrogen, antioxygen, and anticarbon may combine into complex organic compounds," he wrote (Alfvén, 1966, p. 32). "Together with antinitrogen and several other elements they can form the whole range of chemical elements which are the bearers of organic life." Alfvén admitted that it sounded like science fiction, but he insisted that it was a valid scientific prediction based on fundamental physics.

Within the perspective of standard big-bang cosmology, the missing matter problem is a consequence of the exact symmetry between primordially created nucleons and antinucleons. Perhaps the fundamental conservation law that governs the symmetry was not strictly satisfied in the very early and very hot phase of the big bang? This solution to the problem was originally proposed in 1966 by the Russian physicist Andrei Sakharov, who may today be better known as a political dissident and recipient *in absentia* of the Nobel peace prize. Ideas somewhat similar to Sakharov's later became part of the unified theories beyond the standard model known as GUT, an acronym for grand unified theory. Some of these theories allow an initially charge-symmetric state to evolve into a state with an excess of nucleons over antinucleons, producing a universe like the one observed. However, attempts to explain the missing matter problem in this way are not generally considered quite convincing, and so physicists continue to search for other explanations. One of the more fanciful is the antiuniverse before the big bang mentioned earlier. As yet, there is no consensus theory to explain the drastic dominance of matter over antimatter.

.1.4

Constants of Nature

1.4

Constants of Nature

The fundamental laws of nature are associated with and can be expressed by certain parameters the values of which are only known from experiments. As far as is known, the parameters are constant in time and they are therefore generally referred to as constants of nature or sometimes constants of physics. Examples are the constant of gravitation G appearing in Newton's law and the speed of light c appearing in Einstein's theory of relativity. These constants are fundamental in the sense that they do not follow from theory and cannot be derived from other constants (if they could, they would not be truly fundamental). Although it may seem like a tautology to claim that the constants of nature, such as Newton's G or Einstein's c, are independent of time, they are not so by necessity. Some of them may conceivably vary in time, a possibility which has been investigated for nearly a century and which is still being discussed. The term "varying constant" is *per se* an oxymoron, but if the constant is a constant of nature, it is not.

Given the supreme status of the natural constants in modern physics, it is surprising that their importance was only recognized in the last quarter of the nineteenth century. Newton did not state his universal law of gravity in terms of the constant G, which only attracted attention after about 1870, and it is also in this period we have the first measurements specifically of the gravitational constant.

There are many constants of nature, some of them more fundamental than others. Again gravitation provides an example. We are all familiar with the acceleration of gravity at the surface of the Earth, g = ca. 9.8 cm/s^2, but this useful quantity is only relevant to us earthlings and therefore far from fundamental. Had we been Martians, we would have found the surface gravity to be 3.7 cm/s^2, and if Jovians 24.8 cm/s^2. As Newton was aware, all these local constants can be expressed by the same constant G, namely by the relation $g = GM/R^2$, where M is the mass of the planet and R its radius. We also note the unsurprising fact that the numerical values of our constants of nature depend on the chosen system of units. Planck's quantum constant h has the approximate value 6.6×10^{-34} only if expressed in J s (joule times second). In the 1950s, when the unit for energy was "erg," the same constant was ascribed the value 6.6×10^{-27}.

If we want a fixed number, one that does not depend on our conventions regarding units, we need to combine some of the dimensional natural constants into a dimensionless one, a pure number. The masses of the proton and the electron, both of them belonging to the exclusive fundamental class, may be stated in g or kg or in some other mass unit. But the ratio between them, M/m = ca. 1836, is just a number and the same whatever the units. Many physicists believe that the dimensionless constants play a particularly important role in the fabric of nature because they are independent of our conventions.

1.4.1. FUNDAMENTAL CONSTANTS OF NATURE

Perhaps more than other scientists in the pre-World War II period, Arthur Eddington highlighted the significance of the constants of nature. In a book of 1935, he wrote that "We may look on the universe as a symphony played on seven primitive constants as music played on the seven notes of a scale" (Eddington, 1935, p. 227). Of the seven notes or fundamental constants Eddington referred to, six are still to be found in tables of the constants of nature. Apart from the gravitational constant G and the speed of light c, the list comprises Planck's constant h, the elementary charge e, the mass of the proton M, and the mass of the electron m. Eddington included the cosmological constant Λ as the seventh of his notes, which at the time was highly unusual. By the 1930s, the cosmological constant originally introduced by Einstein in 1917 was generally thought to be a mistake, meaning that $\Lambda = 0$. Eddington emphatically disagreed. To him, the constant was no less indispensable than the other constants. Today, after Λ has returned to cosmology, many physicists and astronomers agree with Eddington's view.

Whether six or seven, the values of these constants are given by nature, so to speak, and they cannot be deduced from physical theory. Incidentally, Eddington believed that he could and actually went along calculating theoretical values in striking — perhaps too striking — agreement with those obtained from measurements. For example, for the proton's mass he calculated $M = 1.67277 \times 10^{-24}$ g and the observed value was $M = 1.67248 \times 10^{-24}$ g. However, other contemporary physicists disbelieved his deductions, which they found to be obscure and based on invalid assumptions. This is still the verdict, and in all likelihood it will remain so. While most physicists support the view of six or seven fundamental constants of nature, there is nothing sacred about these numbers. Some physicists want to restrict the number of truly fundamental constants to just three, typically Planck's constant, the speed of light, and the constant of gravitation. Others believe in more constants.

Consider the elementary charge e, which first turned up in Michael Faraday's electrolytic experiments of the 1830s and by the end of the century was identified with the numerical charge of the electron. This charge is elementary insofar that it is the unit of any electrical charge Q. In other words, electricity wherever and whenever it appears is quantized according to $Q = \pm ne$, where $n = 1, 2, 3, \ldots$. (The quarks residing in nuclear particles have fractional charges, $-\frac{1}{3}e$ or $+\frac{2}{3}e$, but they do not exist freely.) This important law of nature dates from about 1915 when Millikan, in a series of famous and later Nobel Prize-rewarded experiments, measured the electrical charge of small ionized oil drops, concluding that it was a multiple of the electron's charge $e = 1.59 \times 10^{-19}$ C (or as Millikan reported, 4.80×10^{-10} electrostatic units).

Millikan's conclusion was not unexpected, but neither was it trivial or self-evident. In fact, for a period of time, it was challenged by the Austrian physicist Felix Ehrenhaft, whose experiments, no less meticulous and precise than Millikan's, gave an entirely different result. Ehrenhaft found evidence for "subelectrons" — charges smaller than e — and questioned the very existence of an elementary charge. The resulting controversy between the two experimenters ended with Millikan's victory, though without Ehrenhaft conceding defeat. He continued to believe that he had detected subelectrons and that neither his nor Millikan's data justified the existence of a smallest charge.

Millikan also used his data to conclude that, if there were a difference between the positive and negative elementary charges, it must be less than $3 \times 10^{-16} e$. He commented that "such neutrality, if it is actually exact, would seem to preclude the possibility of explaining gravitation as a result of electrostatic forces of any kind" (Millikan, 1917). Yet, even such a tiny charge excess might have important macroscopic consequences, such as Einstein speculated at a meeting in 1924, where he discussed the possibility that the proton's charge might exceed that of the electron with the tiny amount $3 \times 10^{-19} e$. Einstein thought that the hypothesis might lead to an explanation of the magnetic fields of the Earth and the Sun.

More than thirty years later, the hypothesis of proton–electron charge inequality entered cosmology in the form of a short-lived but interesting theory aimed at explaining Hubble's expansion law in terms of electrostatic repulsion. Hermann Bondi and his collaborator Raymond Lyttleton argued in a paper of 1959 that a charge excess of $3 \times 10^{-16} e$ — by accident the same as mentioned by Millikan in 1917 — might reproduce the Hubble law and thus make it independent of the equations based on general relativity. The idea was supported by Fred Hoyle, who developed it in his own way. However, within a year the hypothesis was shot down by laboratory measurements showing that the charge excess was smaller than $10^{-19} e$, and consequently Bondi, Lyttleton, and Hoyle were forced to abandon their models of an electrical universe. The status of the charge inequality hypothesis has since changed, as it is now thought to be ruled out theoretically. According to the standard model of elementary particles, the numerical charges of the electron and the proton are exactly the same. The hydrogen atom has no net charge at all, and the same is the case for the neutron.

The status of a constant of nature is not given in advance but varies with the state of knowledge and, therefore, with the historical development. As mentioned in Section 1.2.1, the spectroscopic

Rydberg constant was originally considered a quantity of deep significance, but with Bohr's atomic theory it became degraded to just a combination of other and more fundamental constants. Bohr operated with five of Eddington's constants (e, m, M, c, h) but not with G, which was generally considered foreign to quantum and atomic theory. And, of course, he also did not refer to Λ, the cosmological constant. Somewhat similarly, Avogadro's number, giving the number of particles in one mole of a substance, $N_A = 6.023 \times 10^{23}$, is of great importance in chemistry but not truly fundamental. It can be stated in terms of the more fundamental Boltzmann constant by $N_A = k_B/R$, where R is the gas constant appearing in the equation of an ideal gas.

In his seminal papers of 1899–1900 leading to the hypothesis of energy quantization and the quantum of action, Planck was intensely occupied with the constants of nature and the relations between them. He proposed a set of universal units for length, time, mass, and temperature, later known as Planck units, by combining the fundamental constants of nature. Thus, with modern values he stated the Planck length as

$$l_P = \sqrt{\frac{Gh}{c^3}} = 4.13 \times 10^{-35}\ \mathrm{m},$$

and the Planck time as

$$t_P = \sqrt{\frac{Gh}{c^5}} = 1.38 \times 10^{-43}\ \mathrm{s}.$$

Whereas these units are immensely small, the Planck mass $m_P = \sqrt{ch/G} = 5.56 \times 10^{-8}$ kg is immensely large if compared with the mass of a proton or an electron. From the elementary Planck units, one can construct corresponding units for volume, density, force, and other quantities. The Planck volume l_P^3 is of the order $10^{-105}\ \mathrm{m}^3$ and the Planck density m_P/l_P^3 of the order $10^{97}\ \mathrm{kg/m}^3$; the Planck energy $m_P c^2$ is approximately 10^{31} eV or 10^9 J. Obviously, these units are not suited for everyday purposes.

What appealed greatly to Planck was that his units were de-anthropomorphic, meaning that they were independent not only of specific bodies and circumstances but also of cultural conventions and human norms. They were, he claimed, completely objective. Planck exuberantly wrote that they would be the same for all "extraterrestrials and non-human life forms," should such beings exist. In an address in 1908 on the unity of the physical universe, Planck said about his universal units that they were of such a kind "that the inhabitants of Mars, and indeed all intelligent beings in the universe, must encounter at some time — if they have not already done so" (Planck, 1960, p. 18).

Planck's system of units did not initially attract much attention, except that Eddington in 1918 and at later occasions suggested a very similar system possibly inspired by Planck. However, from a philosophical point of view Eddington's idea of the laws and constants of nature was completely different from Planck's with its emphasis on their detachment from the human mind. For according to Eddington, the constants were numbers introduced by our subjective outlook and for this reason they could be calculated by pure reasoning, without turning up the answer in the book of nature. While Planck wanted to get rid of all anthropomorphic elements, to Eddington such elements were the very foundation of scientific knowledge.

Today, the Planck units or "natural units" as they are also called, are at the heart of foundational physics, not least because they indicate some deep connection between quantum mechanics (h) and general relativity (G). Directly or indirectly, the units are part of modern theories about quantum gravity and about the earliest state of the big-bang universe some 13.8 billion years ago.

1.4.2. MINIMUM LENGTH, MINIMUM TIME

We are used to conceiving space and time as continuous, an assumption on which physics has been securely based since the scientific revolution. Physical objects in space always have a finite size, but space itself can be divided infinitely finely, and the same is the case with time. A physical object may not be ascribed a length of 10^{-1000} m, and periods of time of 10^{-1000} s may not be measurable or associated with physical processes, and yet it makes perfect sense to think about space and time intervals of this order of magnitude. However, the continuity of space-time, or space and time separately, is not self-evident, and through history it has often been questioned. Since the ancient Greeks, philosophers have contemplated the idea that space and time are discrete, consisting of minimal parts in the form of space atoms and time atoms. What for two millennia was a philosophical or theological discourse entered physics in the first half of the twentieth history, and it is still being seriously discussed in some areas of theoretical physics.

With the acceptance of quantum theory and its fundamental postulate of energy quantization, a few physicists wondered if time too might be quantized. After all, energy and frequency are related as $E = h\nu$, so if energy is quantized it is tempting to think also of frequency and therefore time (the inverse of frequency) as consisting of discrete quanta. Considerations of this kind led, in 1926, a French physicist, Robert Lévi, to suggest that time is made up of smallest units of the order 10^{-24} s. For the hypothetical unit, he proposed the appropriate name "chronon" in analogy with, for example, electron and proton. Similar ideas were entertained by other physicists who thought that the hypothesis of time atoms might help explaining some of the problems that faced physics, such as the poorly understood energy spectrum of cosmic rays. The Russian physicist Georgii Pokrowski used the chronon hypothesis to argue for an upper limit of energy density and a maximum frequency of about 10^{23} Hz, for which he claimed support in measurements of cosmic rays.

Other physicists used Heisenberg's uncertainty relations to suggest minimum intervals of space and time. The quantum uncertainty in momentum Δp is related to the position uncertainty Δq by $\Delta p \Delta q \cong h$, which implies that a precise determination of q ($\Delta q = 0$) is possible, but only at the expense of a completely undetermined p. The energy–time relation $\Delta E \Delta t \cong h$ similarly allows for a minimum time interval. Based on such reasoning, physicists around 1930 proposed smallest lengths and durations corresponding to the absolute uncertainties. However, these intervals, which were typically stated as $\Delta q = h/mc$ and $\Delta t = h/mc^2$, depended on whether the mass m referred to an electron or a proton. Space and time atoms of this kind attracted, for a while, the interest of leading quantum physicists such as Heisenberg and Schrödinger, but the interest was short-lived since the hypothesis had little explanatory and predictive force. The chronon or time atom hypothesis failed in particular to lead to progress in physics, and by 1940 it was largely abandoned.

By the early 1930s, quantum physics was widely perceived to be in a state of crisis, principally because of the appearance of divergent integrals (infinities) in the fundamental equations. The grave situation called for drastic cures, one of which was to re-introduce the notion of a universal smallest length, such as Heisenberg suggested in a letter to Bohr of March 1930. Heisenberg's radical idea was to conceive space as a "lattice world" of cubic cells with a size of $(h/Mc)^3$, where M is the proton mass. Realizing the speculative nature of his idea, Heisenberg ended his letter to Bohr: "I do not know if you find this radical attempt completely mad. But I have the feeling that nuclear physics is not to be had more cheaply. Of course I do not take this particular lattice model very seriously" (Carazza and Kragh, 1995). In fact, for a while Heisenberg did take it seriously, and that even though his sketch of a theory violated relativity theory and also the fundamental conservation laws of energy and electrical charge. After Bohr had criticized the idea of a lattice world, Heisenberg decided to shelve it. He did not publish the theory, which is only known from his letter to Bohr and from his correspondence with other physicists.

Still, the general concept of a universal length remained in Heisenberg's mind, and he returned to it on several later occasions. At one stage, he briefly considered to include the constant of gravitation in the form of the Planck length $\sqrt{hG/c^3}$, but since gravity plays no role in particle physics he did not pursue the idea. In 1943, in an ambitious attempt to formulate a unitary theory of all known elementary particles, Heisenberg wrote: "The known divergence problems in the theory of elementary particles indicate that the future theory will contain in its foundation a universal constant of the dimension of a length. ... The difficulty, which still stands in the way ... must probably be perceived as an expression of the fact that, in a manner of speaking, a new universal constant of the dimension of a length plays a decisive role" (Heisenberg, 1943). Nothing important came out of Heisenberg's and others' attempts to create a revised quantum-mechanical theory based on the hypothesis of discrete space or space-time. Nonetheless, the hypothesis would not die, and many years after Heisenberg's death in 1976 it reappeared in modern theories of quantum gravity.

Superstring theory is the best known and most developed unified theory of quantum mechanics and general relativity, but there are several other candidates for a theory of quantum gravity. So-called loop quantum gravity (LQG), an alternative approach developed since the late 1980s, differs in a number of ways from string theory. For example, there are no extra space-time dimensions in LQG and also no supersymmetry. According to LQG, space is not continuous on a very small scale but has a discrete structure somewhat similar to Heisenberg's dream of a lattice world. It follows from the theory that space itself is made up of discrete units characterized by a smallest length of the same magnitude as the Planck length. The volume of a LQG space atom is thus of the order 10^{-105} m^3. Lee Smolin, one of the pioneers of LQG, puts it in this way: "The theory of loop quantum gravity predicts that space is like atoms: there is a discrete set of numbers that the volume-measuring experiment can return. Volume comes in distinct pieces. ... The quantum of volume is so tiny that there are more such quanta in a cubic centimeter than there are cubic centimeters in the visible universe" (Smolin, 2004).

Are there any ways to test experimentally the strange picture of a discrete space? There is, for if space is continuous and Einstein was right, all photons should travel with exactly the same speed irrespective of their energy or frequency. On the other hand, LQG predicts that the speed of light in empty space will depend minutely on the frequency, so that a blue photon moves a bit more slowly than a red photon. The prediction is within the range of experiments. Astronomical measurements made in 2009 of light traveling over more than 7 billion light years showed no difference in speed, and later experiments have also failed to detect the predicted color effect. Although the question is not yet finally settled, the consensus view among experts is that space is continuous and not discrete. Most likely, we do not live in a lattice world.

1.4.3. GRAVITY

The earliest of the presently accepted fundamental constants, the constant of gravitation, was not only at the heart of Newton's celebrated system of the world, it was also of crucial significance to Einstein's very different system. The gravitational constant typically enters in general relativity as the "Einstein constant" κ, which is related to Newton's G as $\kappa = 8\pi G/c^2$. The constant is notoriously difficult to measure, and even today it is not known nearly as accurately as the other constants of nature. By 1930, its accepted value was $G = 6.670 \times 10^{-11}\ \text{m}^3/\text{kgs}^2$, with a relative uncertainty as high as 0.1%. Today it is better known, but not much better.

Inspired by Eddington's ideas about dimensionless combinations of natural constants, in 1937 Dirac considered two such very large constants of the order 10^{39}. The numbers were

$$\frac{e^2}{GmM} \quad \text{and} \quad \frac{T_0}{e^2/mc^3}.$$

In the second expression, T_0 denotes the present value of the inverse Hubble constant $1/H_0$, which is approximately the same as the age of the big-bang universe. Despite its name, the Hubble constant (or parameter) thus depends on time, decreasing slowly as the universe gets bigger and older. The quantity in the denominator is what Dirac called an "atomic time unit," in fact the same as the chronon introduced by Lévi in 1926. Dirac stated as a general principle what later became known as the Large Numbers Hypothesis or LNH: When two numbers of the order 10^{39} or its square 10^{78} appear in nature, they must be related in a simple way; if the numbers are both of the order 10^{39}, they are approximately proportional.

Arbitrarily assuming e, m, M, and c to be true constants, Dirac concluded from the LNH that the constant of gravitation is not really a constant but a parameter slowly decreasing with the age of the universe according to $G \sim 1/t$. The rate of decrease would depend on the Hubble constant, which at the time was not well known. For the relative rate of change and by using $1/H_0 = 2 \times 10^9$ years, Dirac suggested that

$$\frac{1}{G}\frac{dG}{dt} = -3H_0 = \text{ca.}10^{-10} \text{ per year.}$$

Dirac's highly unorthodox $G(t)$ hypothesis was received with surprise and reservation. And understandably so, for it violated Einstein's authoritative theory of general relativity, according to which G is constant. Even more seriously, the hypothesis led to an age of the universe of just about 700 million years — considerably smaller than the age of the Earth! This might seem reason enough to discard Dirac's idea, but at the time also other cosmological models led to an absurdly low age of the universe, so the "age paradox" was not peculiar to Dirac's theory (see Chapter 4.1).

During the 1960s, Dirac and George Gamow discussed in a series of letters the $G(t)$ hypothesis and the possibility of varying constants of nature in general. While Dirac was optimistic, Gamow was skeptical and came up with various empirical objections, such as the too hot climate in the past. In his last letter of 1968, written shortly after Gamow had passed away (unbeknownst to Dirac, of course), Dirac defended his pet idea as follows: "I do not think that the arguments against a varying G are valid. If G varies, then other 'constants' may also be varying, such as the ratio $M_{\text{proton}}/M_{\text{electron}}$ or the coupling constants of nuclear theory. One then cannot build any reliable models of the universe and arguments depending on the usual models are not valid" (Kragh, 1990, p. 238).

With the notable exception of Pascual Jordan, an eminent German theorist and cofounder of quantum mechanics, no leading physicists or astronomers accepted Dirac's claim of a varying gravitational constant. Eddington dismissed it as a fantastic speculation, and to other critics it smelled too much of numerology to be taken seriously. There were even those who accused Dirac of having passed the line separating science from pseudoscience. It did not help when the Hungarian-American nuclear physicist Edward Teller in a paper of 1948 argued that Dirac's hypothesis flatly contradicted established paleontological knowledge. As Teller reasoned, the surface temperature of the Earth depends on the Sun's output of heat energy, a quantity that in turns depends critically on the value of G. Teller calculated that if Dirac were right, the surface temperature of the Earth in the Cambrian era some 250 million years ago would have been close to 100°C. Given the knowledge that the oceans of that era were filled with a rich variety of life, this evidently was a problem for the assumption of a stronger gravity in the past. Teller's objection was not waterproof, but it weakened the credibility of Dirac's hypothesis.

Undisturbed by the methodological criticism and the obvious empirical problems related to his $G(t)$ hypothesis, Dirac maintained his belief in it and the LNH. Without elaborating, in an address of 1939 on the relationship of physics and mathematics, he suggested that the time-variation was a general feature of the basic laws of nature and not restricted to gravity: "At the beginning of time the laws of Nature were probably very different from what they are now. Thus we should consider the laws of Nature as continually changing with the epoch, instead of as holding uniformly throughout space-time" (Dirac, 1939). And this was not all, for he continued to speculate that the laws and constants of nature might also vary in space: "We should expect them also to depend on position in space, in order to preserve the beautiful idea of the theory of relativity that there is a fundamental similarity between space and time." It is unclear how seriously Dirac meant the proposal and if he really thought that the laws of nature in some distant galaxy differ from those known from our region of the universe.

Dirac only returned to the LNH and the associated $G(t)$ hypothesis in 1972, but in the meantime hypotheses of a varying G were proposed by other physicists. Carl Brans and Robert Dicke, two physicists at Princeton University, proposed in 1961 a new theory of gravitation, which they announced as an extension of or alternative to Einstein's theory of general relativity. Inspired by but not relying on Dirac's LNH, the Brans–Dicke theory included a varying gravitational constant, but one which decreased more slowly than in Dirac's case. Instead of a variation given by $G \sim t^{-1}$, Brans and Dicke suggested $G \sim t^{-0.1}$. In a series of works through the 1960s, Dicke investigated in detail the astronomical and geological consequences of a decreasing G. What effect would the hypothesis have on the Earth's climate in the past? How would it affect the motion of the Moon and the luminosity of the Sun? Frustratingly, the efforts that Dicke and other scientists devoted to these questions did not result in a clear answer regarding the varying gravity hypothesis. Evidence from astronomy and paleontology did not support the hypothesis, but neither did it unambiguously rule out a weak decrease of gravity over time.

Jordan too was much occupied with these questions, but contrary to Dicke he favored Dirac's original hypothesis of $G \sim t^{-1}$. In the mid-1950s, Jordan came to the conclusion that, as a result of decreasing gravity, the Earth would have been considerably smaller in the past and since then have expanded to its present size. The idea of an expanding Earth was not new, but Jordan was the first to base it on the assumption of a decreasing G, such as he did in several papers and a monograph of 1971 titled *The Expanding Earth*. Jordan estimated the radius of the Earth to expand by a rate about 5 mm per year, whereas Dicke suggested the much slower expansion rate of $dR/dt = 0.05$ mm per year. For a decade or so, the expanding Earth hypothesis was much discussed among physicists as well as geologists, but it was more than difficult to find empirical evidence that clearly supported the hypothesis. At the time when Jordan published his book, the idea of an expanding Earth was on its way to be marginalized and separated off from mainstream geophysics. Today, it is known that if the Earth expands, it does so inappreciably. The best data for recent decades result in $dR/dt = (0.35 \pm 0.47)$ mm per year, meaning that the expansion rate is essentially zero.

The more fundamental question of $G(t)$ could only be settled by means of precise methods based on astronomical measurements, but also in this case the answers were ambiguous and generally disappointing to advocates of varying gravity. In the early 1970s, radar measurements of the distances to the Moon and the interior planets Mercury and Venus pointed to a relative change in G less than 10^{-10} per year, and even more stringent constraints came from the Viking landers on Mars and the Lunar Laser Ranging Project (LLRP). In the latter project, a reflector placed on the Moon in 1969 provided the scientists with thousands of precise data on the Earth–Moon distance. A study of 2007 based on the accumulated LLRP data concluded that

$$\frac{1}{G}\frac{dG}{dt} = (2 \pm 7) \times 10^{-13} \text{ per year,}$$

which corresponds to less than a 1% variation of G over the 13.8 billion age of the universe. Thus, the gravitational constant is in fact constant, in agreement with the standard Einstein theory of general relativity.

On the other hand, experiments can never prove a physical quantity to be absolutely constant, but only that the time variation is smaller than a certain number given by the experimental uncertainty. Perhaps surprisingly, the verdict from the astronomical measurements has not put the $G(t)$ hypothesis in the grave. The consequences of the hypothesis, in one or other of its several versions, are still explored by a minority of physicists in the hope that it will lead to a better understanding of problems in cosmology, particle physics, and gravitation theory. For example, some physicists have tried to explain the mysterious content of dark matter in the universe on the basis of varying-G theories. Although Dirac's version has been definitely refuted, his predilection for large dimensionless numbers and varying constants is shared by more than a few modern physicists.

1.4.4. THE FINE-STRUCTURE CONSTANT

Among the dimensionless constants of nature, no one is as famous and magical as the fine-structure constant defined as

$$\alpha = \frac{e^2}{\hbar c} = \frac{2\pi e^2}{hc}.$$

In numerical terms, α = ca. 0.0073 and for the inverse quantity $1/\alpha$ = ca. 137.036. From a modern point of view, the α constant is one of four coupling constants that characterize the relative strengths of the four fundamental interactions, in this case the electromagnetic interaction between charged particles (electrons) and the electromagnetic field (photons). Expressed in Planck units, where $c = \hbar = 1$, the fine-structure constant becomes $\alpha = e^2$ and the denominator $\hbar c$ is just a scale factor.

From a historical perspective, the fine-structure constant has two roots. First, it was realized from an early date that the quantum constant h has approximately the same order of magnitude as e^2/c, and that the dimensionless ratio e^2/hc might in some way be interesting. Planck and Einstein were the first to point out the possibility, which they did around 1908. The second and quite different root of what became the fine-structure constant turned up in spectroscopy in the late 1880s, when Albert Michelson and other physicists discovered that the bright red line in the Balmer spectrum was a doublet consisting of two closely spaced lines. The hydrogen lines, not only the red one but also the other lines, have a "fine-structure." When the German physicist Arnold Sommerfeld in an important work of 1916 analyzed the hydrogen atom theoretically, which he did by introducing relativity theory in Bohr's atomic model, the two roots merged into one.

In a hydrogen atom in its lowest quantum states, the electron moves in elliptical orbits at such a high speed that its mass increases in accordance with the theory of relativity. By taking this into consideration, Sommerfeld could calculate the exact energy levels from which the fine-structure of the spectral lines emerged in beautiful agreement with experiments. Sommerfeld named the quantity $2\pi e^2/hc$ appearing in his formula the fine-structure constant, reporting its value to $\alpha = 0.00726$ (at the time he did not pay attention to the inverse value). Although appearing merely as a spectroscopic quantity, Sommerfeld thought that α had a wider significance and that it suggested some deep and as yet mysterious connection between electrodynamics (e), relativity (c), and quantum theory (h).

In 1928, Dirac reproduced Sommerfeld's fine-structure formula on the basis of his new relativistic wave equation, a result which increased the status of α as a most fundamental constant of nature (Section 1.3.2). What was the true meaning of α? Why did it have the particular value given by experiments, rather than some other value? From that time onwards, theorists became fascinated — some obsessed — with the problem of deriving the fine-structure constant from fundamental theory. Referring to $1/\alpha$ = ca. 137, Richard Feynman described the enduring fascination of the fine-structure constant as follows: "It has been a mystery ever since it was discovered more than fifty years ago, and all good physicists put this number up on their wall and worry about it. … It's one of the *greatest* damn mysteries of physics: a *magic number* that comes to us with no understanding by man. You might say that 'the hand of God' wrote that number, and 'we don't know how He pushed His pencil'" (Feynman, 1985, p. 129).

Eddington would not have agreed with Feynman's quote — but then, perhaps Feynman did not count Eddington as belonging to the "good physicists." It was only with the unorthodox research program of the British astronomer-physicist that the fine-structure constant was elevated from an empirical quantity to a truly fundamental one. From his own interpretation of Dirac's wave equation, Eddington derived in 1929 that $1/\alpha = 136$ exactly (confusingly, he always used the name fine-structure constant and the symbol α for $hc/2\pi e^2$ and not for its inverse). Eddington insisted that $1/\alpha$ must be a whole number, and when experiments consistently showed that it was close to 137, he found reasons for adding one unit. Unmoved by the experimenters' value, which around 1940 was $1/\alpha = 137.030 \pm 0.016$, he stuck to the number 137. In a paper of 1932 Eddington concluded that "the value 137 is here obtained by pure deduction, employing only hypotheses already accepted as fundamental in wave mechanics" (Eddington, 1932). Bohr, Pauli, Dirac, and most other experts in quantum mechanics disagreed.

Few mainstream physicists believed in Eddington's theoretical value, but it was not without impact. To mention but one example, although Max Born sharply criticized Eddington's aprioristic methodology, he too was captivated by the possibility of deriving theoretically the value of the fine-structure constant. One of Born's papers from the mid-1930s carried the title "The Mysterious Number 137." Moreover, like Eddington, he believed that the fine-structure constant, or the approximate number 137, was closely connected with another pure number, the proton–electron mass ratio M/m = ca. 1840. Born's elaborate calculations resulted in a rough relationship between the two constants and even in an order-of-magnitude value of $1/\alpha$, but he eventually came to the conclusion that his work in this area was wasted, nothing but physically useless mathematics. Attempts to derive the fine-structure constant theoretically or by means of numerological reasoning — what pejoratively has been called "alpharology" — have continued to this day. As a mathematician pointed out in 1969,

$$\frac{9}{8\pi^4}\left(\frac{\pi^5}{2^4 \times 5!}\right)^{1/4} = \frac{1}{137.036082\ldots}$$

a value surprisingly close to the measured fine-structure constant. This is perhaps interesting, but of course it is not an explanation.

During the 1930s, after the discovery of the expanding universe, several scientists attempted to explain the galactic redshift-distance data on the assumption that the universe was after all static (see Section 3.4.5). It was in this context that two British physicists in 1935 introduced the idea of a varying fine-structure constant, which they did by assuming e^2 to vary linearly with cosmic time. A similar idea was considered by Gamow in 1967. As Gamow pointed out, the hypothesis of $e^2 \sim t$ or $\alpha \sim t$ implies Rydberg's spectral constant to vary as t^2, the reason being (Section 1.2.1) that the latter constant can be written as $R = \alpha^2 mc^2/2h$. Gamow believed that the variation might have testable effects with regard to galactic spectra. Although nothing came out of Gamow's speculation, the idea survived that measurements of the fine-structure doublets from objects far away might reveal a difference depending on the age of the objects.

Based on a study of absorption lines from distant quasars, the Australian astrophysicist John Webb and his collaborators reported in 2001 evidence that the fine-structure constant was smaller in the cosmic past. The relative variation was small but significant, about 5×10^{-16} per year. The announcement caused great excitement and was eagerly followed up by physicists and astronomers. Alas, new and more precise data released in 2004 showed no variation in the fine-structure constant of light received from quasars. Nor has $\alpha(t)$ turned up in later observations. As recently as April 2020, a team headed by Webb reported results from an analysis of light emitted by quasars formed only 800 million years after the big bang. The conclusion of the study was that "the weighted mean value of α is consistent with the terrestrial value and is $\Delta\alpha/\alpha = -2.18 \pm 7.27 \times 10^{-5}$" (Wilczynska *et al.*, 2020). Thus, the situation with respect to $\alpha(t)$ is today more or less the same as it is with respect to $G(t)$. Although a tiny variation of α cannot be ruled out, there is no convincing experimental evidence that it is not what its name indicates, a true constant of nature.

In analogy with α being the coupling constant of electromagnetic interactions, there are dimensionless coupling constants for the strong and weak forces described by the standard model of particle physics. For example, the coupling constant for the strong force is about 137 times larger than α. According to the standard model, at the extremely high energy or temperature of the very early universe, about the time 10^{-35} s after the big bang, the three forces had the same strength or approximately so. As the universe expanded and cooled, α decreased from its original value of perhaps 1/40, ending up with its present value of about 1/137. What supposedly happened in the early universe has received support from high-energy experiments showing that the value of α depends on the energies involved. Thus, the α constant has its present value only because we live in a cold universe. Since the energy and temperature of the universe is a measure of its age, there is a sense in which the fine-structure constant depends on cosmic time. However, this variation is limited to the very early universe and is entirely different from what is normally referred to as a time-dependent fine-structure constant.

1.4.5. THE VELOCITY OF LIGHT

In his masterpiece *Principia* of 1687, Newton referred to the Danish astronomer Ole Rømer's discovery eleven years earlier that light moves with a finite speed rather than propagating instantaneously as traditionally believed. After Newton's compatriot, the priest and natural philosopher James Bradley, had established in 1728 that all stars show the same aberration, it was realized that light propagates with the same speed throughout the universe. Expressed in our units, Bradley determined the new universal constant of nature to be $c = 301{,}000\,\text{km/s}$. However, it was only with Maxwell's electromagnetic theory of light and Einstein's theory of relativity that the quantity c came to be seen as truly fundamental.

As Einstein postulated in his revolutionary paper of 1905, the speed of light in vacuum is independent of the motion of the light source. In addition to this counterintuitive claim, he stated that "velocities greater than that of light have ... no possibility of existence" (Einstein *et al.*, 1952, p. 64). A similar claim had been made by Henri Poincaré one year earlier, but it was only with Einstein's theory that it became commonly accepted. In fact, Einstein was mistaken in his original belief that superluminal velocities are incompatible with the theory of relativity. As first demonstrated by the Russian physicist Lev Strum in the mid-1920s, hypothetical faster-than-light particles (later known as tachyons) do not contradict special relativity insofar that they always move with a speed greater than c. This speed is an impenetrable barrier for ordinary particles with $v < c$ and also for tachyons, if the existed, with $v > c$. There have been many searches for tachyons, but none of them have provided any evidence that they exist.

The velocity of light has always been difficult to measure, such as illustrated by Michelson's authoritative value of 1926, which was $c = 299\,796 \pm 4\,\text{km/s}$. Today, it makes no sense to improve the value of c by means of still more advanced measuring instruments. The reason is that since 1983, c has been *defined* to have the exact value 299 792.458 km/s and therefore with no attached experimental uncertainty. Likewise, in 1983 the meter was redefined as the distance traveled by light during $1/c$ of a second. While G and α are measurable constants, the status of c is different. It is often regarded as just a factor converting length in meters to time in seconds.

Whether before or after 1983, many physicists and astronomers have discussed the possibility of c varying in time, just as they have done with other constants of nature. In a paper of 1911, Einstein argued that speed of light would depend on the gravitational field, which in the case of a static field means a variation with position rather than time. The first suggestions of a $c(t)$ variation date from about 1930, when a few scientists speculated that the velocity of light might systematically decrease in time. Among the proposals was the linear relation $c(t) = c_0(1 - kt)$ with k a frequency constant. The only justification for a time variation of this kind was that it, with an appropriate value of k, led to an explanation of the galactic redshifts without assuming an expanding universe (Section 3.4.5). The proposals attracted limited attention, but they annoyed Eddington, who was convinced that the constants of nature could not possibly vary. He argued that since c was given by the ultimate standards of length and time, it must be the same everywhere and at any time. As Eddington dismissed Dirac's $G(t)$ hypothesis as "fantastic," so he dismissed the $c(t)$ speculations of the 1930s with words such as "nonsense" and "self-contradictory." For half a century, hypotheses of varying speed of light were considered speculative and scarcely taken seriously.

When $c(t)$ hypotheses returned to respected scientific journals in the 1990s, it was in a cosmological context and primarily with the aim of constructing models of the early universe that were alternatives to the popular but controversial inflation theory

(Section 4.2.4). According to the inflation theory or scenario dating from about 1980, the big bang initially resulted in a brief vacuum state expanding at a phenomenal rate. During this phase of only 10^{-30} s or so, space inflated by the unbelievable factor of about 10^{40}! What is known as VSL cosmology (Varying Speed of Light) took off with a paper of 1999 written by João Magueijo and Andreas Albrecht. Rather than assuming an early inflation phase, the two authors assumed the speed of light to have decreased drastically at a time close to the Planck era and after this drastic decrease to have settled on its current value. This first example of VSL cosmology was soon followed by a proliferation of models similarly based on the assumption of a varying speed of light in the past.

A variety of VSL theories are still being explored, but although this kind of theory is in some quarters considered an interesting alternative to inflation, it has not succeeded in replacing inflation as cosmologists' favored scenario of the very early universe. The standard view is that the speed of light is constant, indeed must be constant, and VSL theories have not seriously changed this view. The critical responses to VSL theory are no less interesting than the theory itself, for other reasons because they illustrate how ideas of a conceptual and philosophical nature unavoidably turn up in fundamental physics. One of these ideas is conventionalism, the belief that our knowledge of nature (or society) principally rests on human conventions and not or only in part on an independent and objective nature.

Elements of conventionalism are central to the discussion of VSL versus standard cosmology. After all, if c is chosen by convention as a conversion factor between length and time, as it has been since 1983, how can it possibly change? It is generally agreed that whereas the variation of dimensionless quantities, such as the fine-structure constant, can be determined by observation, c and other dimensional constants are not given by nature but by human conventions. According to Michael Duff, a critic of VSL, the question of a varying c is no deeper than the question of whether the number of liters to a gallon varies or not. Echoing Eddington's old complaint, Duff writes: "There is no such thing as a varying c 'theory' only varying c 'units.' For example, in units where time is measured in years and distances in light-years, $c = 1$ for ever and ever, whatever your theory!" (Duff, 2015).

Part 2

Molecules and Elements

.2.1

Inorganic Darwinism and the Unity of Matter

2.1

Inorganic Darwinism and the Unity of Matter

Since the late eighteenth century, chemistry has been firmly based on the concept of elements traditionally thought to be the distinct and immutable building blocks of all matter. In 1800, the number of recognized elements was about 40, and a century later the number had doubled; today, we have knowledge of 118 different chemical elements (Chapter 2.5). The dream of reducing the number of elements to just one fundamental substance, a particle or something else, played an important if controversial role throughout nineteenth-century chemistry, and the dream was shared by more than a few physicists and astronomers. Although it belonged more to chemical philosophy than to mainstream laboratory chemistry, it was no less important for that. If there were a primitive form of matter from which the known elements had somehow evolved, perhaps it was to be found in the stars rather than in terrestrial matter. Since the 1860s, spectroscopy opened up for the study of hypothetical elements belonging to the heavenly regions, a possibility that a minority of chemists associated with the new periodic classification of the elements. Darwin's sensational theory of organic evolution further stimulated speculations of a fundamental unity in nature, the idea that the chemical elements had slowly evolved in a manner similar to living species. Could the evolution be traced back to a common ancestor of all elements?

Although the search for unity and progressive evolution was bound to be speculative, at the end of the nineteenth century it appeared that it was more than just a philosophical dream. Radioactivity questioned the fixity of elements and the electron more than suggested that atoms are composite bodies. By and large, the speculations of the more philosophically minded chemists in the Victorian era proved immensely productive. One of the results was the surprising emergence of "astrochemistry," a new interdisciplinary field that not only changed the scope of chemistry but also, and to an even greater degree, that of astronomy. And it went further, for the speculations also involved the beginning of the material universe as a whole, that is, scenarios of a cosmological nature. Although "cosmochemistry" only became a reality in the 1950s, its seeds were planted in the second half of the nineteenth century.

2.1.1. PROUT'S LAW

Chemical atomism in a form recognizable today was introduced by the English schoolteacher and natural philosopher John Dalton, who in his innovative textbook of 1808, *A New System of Chemical Philosophy*, associated chemical elements with particular atoms. Dalton's atoms were indivisible and permanent elementary bodies, and the atoms of the same element were identical whether the element was pure or part of a compound. Most importantly, he introduced the crucial notion of a relative atomic weight, thereby associating an element with a measurable quantity. It followed that there must be as many elements as there are atomic species and that a pure element can be defined as a substance with a particular atomic weight. Using hydrogen as a unit, in *A New System* Dalton found from measurements and certain simplicity assumptions that the atomic weight of oxygen was about 7 and the one of sulfur about 13. He stated that an "atom of water" consisted of one hydrogen atom combined with one oxygen atom, that is, HO rather than the later accepted H_2O. His stoichiometric formula for an "atom of acetic acid" corresponded to $C_2H_2O_2$.

Apart from the crucial element-atom association, Dalton's understanding of an element was empirical and operational, and in this respect similar to the one which Antoine-Laurent Lavoisier had famously stated in his *Traité Élémentaire de Chimie* (Treatise of Elementary Chemistry) from 1789. An element or "simple body" was one that had not been decomposed but could combine with other bodies. Although the reality of Dalton's atoms was frequently doubted during the nineteenth century, the Lavoisier–Dalton definition won general acceptance.

While Dalton was adamant that the atoms of the various elements were different and maximally simple bodies, in 1815 the British medical doctor William Prout suggested that since the known atomic weights were close to integers, improved measurements would probably show the approximate equality to be an exact one. In other words, he suggested that for any element x the atomic weight could be stated as a multiple of hydrogen's weight, that is $A_x = nA_H$, where $n = 1, 2, 3, \ldots$. From this empirical version of "Prout's law," it was tempting to draw the ontological conclusion that all the heavier elements and hence all matter consist of hydrogen atoms. In that case, the hydrogen atom would be the one and only ultimate particle, the long sought *protyle* (as Prout called it), which is indeed what Prout and his followers believed.

For philosophical reasons, the idea of a basic unity of matter as embodied in Prout's hypothesis appealed to many nineteenth-century chemists. But unfortunately, it disagreed with more careful determinations of the atomic weights. Experiments showed that although in most cases the atomic weights were close to a whole number, they nonetheless differed from it, and in a few cases they were far from being integers. Chlorine with an atomic weight of approximately $A = 35.5$ and also copper with a weight of $A = 31.7$ (as given about 1860) squarely contradicted Prout's law or hypothesis. With the very precise atomic weight determinations of the Belgian chemist Jean Servais Stas published in 1860, Prout's law greatly lost its credibility. Stas' measurements proved beyond doubt that, generally, the atomic weights of the elements were not multiples of that of hydrogen and that consequently the original form of Prout's law had to be rejected. Although this conclusion was generally accepted, it did not imply the death of the idea of a material unity underlying the law.

In the eyes of many chemists, the idea was too beautiful to be wrong.

In fact, it was easily possible to modify the idea of an atomic unity of matter into versions that were not refuted, or even not refutable, by experiment. These possibilities were eagerly explored, and much ingenuity was applied in saving the philosophically attractive idea. One way of retaining Prout's law in spite of the measurements of Stas and others was simply to reduce the magnitude of the prime unit until it fitted experimental values. If experiments ruled out hydrogen as the prime unit of matter, why not postulate an even smaller unit, preferably with a weight of a half or a quarter of that of hydrogen? Formally, the modified Prout formula would then be $A_x = \frac{1}{2} n A_H$ and $A_x = \frac{1}{4} n A_H$, or even $A_x = 1/a \, nA_H$ with $a > 4$. The hypothetical unit could be arbitrarily small, thus automatically securing agreement with measurements. In other words, the modified formula was unfalsifiable. The great Swiss chemist Jean Charles Marignac favored such an opportunistic approach. In a paper of 1860, he wrote:

> "The fundamental principle which has led Prout to the enunciation of his law, that is, the idea of the unity of matter ... [is] quite independent of the magnitude of the unit which is to serve as the common divisor for the weights of the elementary atoms, and which can therefore be looked upon as expressing the weight of the atoms of primordial matter. If this weight should prove to be that of one atom of hydrogen, or of a half, a quarter atom of hydrogen, or if it should be ... a hundredth or a thousandth for instance, the same degree of probability would attach to it" (Freund, 1904, p. 601).

While the sub-hydrogen atoms were in most cases purely hypothetical, some chemists believed that the case of $A = \frac{1}{2} A_H$ was particularly interesting because it was embodied in the form of the element helium.

Another saving operation was the so-called ether condensation hypothesis, according to which the deviations from integral atomic weights were due to amounts of the ethereal substance that condensed upon the surface of material atoms. Although the ether was primarily of interest to the physicists as the medium through which light was transmitted, it also played a role in the way that many chemists thought about the secrets of matter. As early as 1872, shortly after having proposed his version of the periodic system, the distinguished German chemist Lothar Meyer imagined that "apart from the particles of this primary matter [hydrogen] there may be included in the constitution of the atom larger or smaller amounts of that substance, perhaps not being completely weightless, which fills up the universe and we use to call the light ether" (Meyer, 1872, p. 293). He repeated the suggestion in later editions of his influential textbook *Die Modernen Theorien der Chemie* (Modern Theories of Chemistry).

The ether condensation hypothesis had the advantage that it could be tested experimentally, given the reasonable assumption that the quantity of condensed ether would change during a reaction between two different substances. For example, in the formation of mercury(II) bromide from its constituents, $Hg + Br_2 \rightarrow HgBr_2$, one would expect a small reduction in weight. Precision experiments in 1891 with the mercury-bromine reaction proved that the relative change of weight was less than 5×10^{-8}, much too small to explain the deviations from Prout's law by means of the hypothetical ether mechanism. A few years later, the German chemist Hans Heinrich Landolt began a series of similar weight determinations of unsurpassed accuracy, originally with the expressed purpose of deciding whether the ether had any place in chemistry. He tested the law of mass conservation in four different reactions, among them the reduction of iodine with sodium thiosulfate:

$$I_2 + 2Na_2S_2O_3 \rightarrow 2NaI + Na_2S_4O_6.$$

Landolt's conclusion was negative and, in 1893, he stated that "the last resource which was still open for Prout's hypothesis is now blocked" (Landolt, 1893). However, supporters of the hypothesis disagreed, among them none other than Mendeleev (Section 2.1.3). Although the weight determinations of Landolt and others ruled out the ether condensation hypothesis, they did not rule out the possibility of an unknown primordial element with atomic weight smaller than that of hydrogen. Prout's hypothesis was still part of the chemists' game.

2.1.2. THE RISE OF ASTROCHEMISTRY

In 1835, the influential French philosopher Auguste Comte, known as the father of positivism, stated confidently in his *Cours de Philosophie Positive* (Course of Positive Philosophy) that the stars could never be subjects of scientific study. Being inaccessible to experiments, the physical and chemical composition of the stars would forever remain unknown. Comte was wrong, completely wrong, for less than thirty years later the new spectrum analysis established by Gustav Robert Kirchhoff and Robert Bunsen enabled chemists, physicists, and astronomers to gain information on what Comte had claimed was impossible. One of Kirchhoff's earliest spectroscopic papers, published in *Philosophical Magazine* in 1861, was entitled "On the Chemical Analysis of the Solar Atmosphere," an indication that the new field was widely seen as associated with chemistry. Astrospectroscopy provided chemistry with a grander perspective and opened up for insights in and speculations about the origin and nature of chemical elements. Do there exist elements in the stars that are not found on the Earth? Do stellar spectra provide evidence of matter in some primordial state and of the complexity of elements? Prout's hypothesis and the new periodic system furnished the theoretical components that for a period made astrochemistry an exciting if speculative branch of interdisciplinary science.

Spectroscopy transformed both chemistry and astronomy and in general brought the two sciences, traditionally thought to be not only different but even incompatible, into closer contact. During the Victorian era, it became common to speak about the stars and galaxies as heavenly laboratories, a metaphor first used by William Herschel in 1785. In a retrospective account of 1897, the prominent amateur astronomer William Huggins noted that "the astronomical observatory began, for the first time, to take on the appearance of a laboratory. … The observatory became a meeting place where terrestrial chemistry was brought into direct touch with celestial chemistry" (Crowe, 1994, p. 185). Other scientists, who were closer to chemistry than Herschel and Huggins, contemplated whether the conditions of the stars could somehow be reproduced in the laboratory.

Although the general view was that all parts of the universe shared a common chemistry, more than a few scientists believed that in some respects the chemistry of the heavens might differ from that of the Earth. Several stellar spectral lines could not be identified with lines known from the laboratory, and based on such scant evidence it was tempting to suggest that they were due to new, non-terrestrial elements (see also Chapter 2.2). Stellar spectroscopy led to speculative predictions of about a dozen elements, which have only in common that they do not exist. While this was the case with coronium, nebulium, asterium, and austriacum, to name just a few, the single and most noteworthy exception was helium. This element turned up as a spectral line in 1868; in 1895, it was recognized to be real; and shortly thereafter, it made its entry into the periodic table.

In spectroscopic studies of a solar prominence in 1868, Norman Lockyer, an accomplished British amateur astronomer and the founder of the journal *Nature*, detected a hitherto unknown spectral line of wavelength 5876 Å which he designated as D_3. Contrary to what is usually stated in the literature, at first he did not think of it as due to a new chemical element and he did not refer to it as a "helium" line. Only during about 1873 did Lockyer, collaborating with the chemist Edward Frankland, cautiously suggest a new solar element. During the following two decades, helium remained a shadowy substance that was often referred to but far from generally accepted. No one dared suggesting a place for it in the periodic system. Since practically nothing was known about helium, the hypothetical solar element invited a variety of speculations, many of them associated with the idea of a primary matter lighter than hydrogen.

The esteemed American chemist Frank Clarke, at the U.S. Geological Survey, was the first to speculate, in a paper of 1873, that helium might have an atomic weight of one half and therefore be the primeval element needed to make the atomic weights compatible with Prout's law. The same idea was suggested by a few other scientists, most visibly by the British chemist William Crookes, who in 1861 had discovered the element thallium by means of the spectroscope. In a wide-ranging address of 1886 to the British Association for the Advancement of Science, Crookes told his audience that, "As D_3 has never been obtained in any other spectrum, it is supposed to belong to a body foreign to our Earth, though existing in abundance in the chromosphere of the sun"

(Crookes, 1886). He went on: "Granting that helium exists, all analogy points to its atomic weight being below that of hydrogen. Here, then, we may have the very element, with atomic weight half that of hydrogen, required by Mr. Clarke as the basis of Prout's law." By arbitrarily assigning $A = 0.5$ to helium, Crookes could accommodate the measured weights of the elements with the $A_x = ½nA_H$ modification of Prout's law.

The imaginative Crookes mentioned yet another way to save the cherished hypothesis, namely, that the atomic composition of the chemical elements might not be uniform but consist of atomic species with different atomic weights. His proposal, as he stated it in the 1886 address, was this: "When we say that the atomic weight of, for instance, calcium is 40, we really express the fact that, while the majority of calcium atoms have an actual atomic weight of 40, there are not a few which are represented by 39 or 41, a less number by 38 or 42, and so on." At the time this was nothing but an airy speculation, but with the discovery of isotopes more than twenty years later, it turned out to be an anticipation of a most important revision of what constitutes a chemical element. As far as helium is concerned, after William Ramsay in 1895 had detected the element in terrestrial minerals, within a year or two its atomic weight was established to be close to 4. It was not, after all, lighter than hydrogen.

Helium was not Lockyer's only contribution to the question of the evolution and complexity of atoms. A leading proponent of the so-called dissociation hypothesis, he was convinced that in the intensely hot stars the chemical elements would decompose to smaller and simpler forms of matter. Not only did the stars evolve, so did the atoms, and the drive behind the evolution was temperature:

> "If the various stars ... bring before us a progression of new forms in an organized sequence, we must regard the chemical substances which visibly exist in the hottest stars which, as far as we know, bring us in presence of temperatures higher than any we can command in our laboratories, as representing the earliest evolutionary forms" (Freund, 1904, p. 618).

Lockyer's straightforward line of reasoning was that if chemical compounds dissociated in a gas burner, at the much higher temperatures in the stars the elements themselves might dissociate. Lockyer first proposed the astrochemical dissociation hypothesis in 1873 and expounded it fully in the later monographs *Chemistry of the Sun* (1887) and *Inorganic Evolution* (1900). His enthusiasm for element dissociation made him experiment with strong electric discharges in the hope that they would cause elements to decompose. In 1879, he created a sensation by reporting that he had actually decomposed sodium, sulfur, and some other elements into hydrogen. Alas, Crookes and other chemists were unable to repeat the experiment and it soon turned out that the hydrogen was due to impurities. Embarrassed, Lockyer retracted the transmutation claim, but not his belief in decomposable elements at very high temperatures.

Without considering the stars, Lockyer's belief was shared by some chemists, notably Victor Meyer, a professor at the University of Heidelberg. In the 1890s, Meyer and his collaborators engaged in a large-scale research program in pyrometry with the aim of investigating molecular and possibly atomic decomposition at temperatures approaching 2000°C. Meyer was convinced that atoms were composite bodies and that they would decompose to simpler particles at sufficiently high temperatures.

2.1.3. PERIODIC SYSTEMS

The periodic system of the elements dates from 1869, when Lothar Meyer in Germany and Dmitri Ivanovich Mendeleev in Russia presented their different but equivalent versions of the system. In both cases, the ordering parameter was the atomic weight A, as it continued to be up to about 1920, when A was replaced by the atomic number Z. Mendeleev was soon recognized as the true founder of the system, in particular because his version was more informative and had greater predictive power than Meyer's. Famously, from vacant places in the system he predicted the existence of "eka-aluminum," "eka-silicon," and "eka-boron," or what came to be known as gallium, germanium, and scandium. As mentioned, Lothar Meyer suspected the atoms to be composite bodies, not indivisible but merely undivided as yet. Mendeleev would have nothing of this heresy, as he thought it was. He also resisted speculations based on Prout's hypothesis and attempts to interpret the periodic system as evidence that the elements had come into being through an evolutionary process. To him, the individuality and permanence of the elements, and with it the fixity of the periodic table, was a dogma without which chemistry would not be a proper science.

Characteristically, Mendeleev dismissed the hypothetical helium as nothing but a chimera, part of an unhealthy trend in chemical philosophy. In his Faraday Lecture of 1889, he said: "As soon as spectrum analysis appears as a new and powerful weapon of chemistry, the idea of a primary matter is immediately attached to it. From all sides we see attempts to constitute from the imaginary substance helium the so much longed for primary matter" (Mendeleev, 1889). According to Mendeleev, the $\lambda = 5876$ Å spectral line was probably due to a known element placed under such extreme conditions of temperature and pressure that the line could not be reproduced in laboratory experiments. Five years later, when Ramsay and Lord Rayleigh controversially claimed the existence of argon, a new monatomic gas with atomic weight close to 39.9, Mendeleev protested that there was no place for it in the periodic system. He consequently suggested that the so-called argon element might in fact be N_3, an allotropic form of nitrogen. It took some years until Mendeleev accepted helium, argon, and the other inert gases (which by 1900 also comprised neon, krypton, and xenon).

Crookes was fascinated by the periodic system, but he conceived it and the elements it comprised quite differently from Mendeleev's view, namely, as an evolutionary scheme. In 1888, he presented a three-dimensional spiral model of the periodic system aimed at illustrating the evolution of the elements from hydrogen to uranium. While Mendeleev and most other chemists believed that the remaining gaps in the system represented bona fide elements waiting to be discovered, according to Crookes' 1886 address mentioned earlier, "these gaps may only mean that at the birth of the elements there was an easy potentiality of the formation of an element which would fit into the place." This view of un-actualized possibilities was unconventional and in sharp contrast to Mendeleev's. No less unconventional was

Crookes' cosmic scenario, again in his 1886 address to the British Association, of how the elements had originally come into existence:

> Let us picture the very beginning of time, before geological ages, before the earth was thrown off from the central nucleus of molten fluid, even before the sun himself had consolidated from the original *protyle*. ... Let us assume that the elementary *protyle* contains within itself the potentiality of every possible combining proportion of atomic weight. Let it be granted that the whole of our known elements were not at this epoch simultaneously created. The easiest formed element, the one most nearly allied to the *protyle* in simplicity, is first born. Hydrogen — or shall we say helium? — of all the known elements the one of simplest structure and lowest atomic weight, is the first to come into being.

Surely, here we have a spirited if purely speculative anticipation of how the chemical elements were formed in nuclear processes in the early phase of the universe (Chapter 3.3).

Not only did Mendeleev ignore or dismiss evolutionary speculations à la Crookes, he also disbelieved the new subatomic physics which was heralded in the late 1890s with the discoveries of radioactivity and the electron. To his mind, the main problem with these theories was that they threatened to ruin the foundation of traditional chemistry based on the dogma of immutable atoms. He seriously believed that if subatomic particles and transmutation of elements were accepted, chemistry would degrade into a pre-scientific state similar to the alchemy of the past. To abandon the immutable atom was not only undesirable, it would have catastrophic consequences for chemistry as a science. In the early years of the new century, Mendeleev thought to have found an alternative to the new physics in the form of a chemical ether, a hypothesis that appeared prominently in the third edition of his massive and widely read textbook *Principles of Chemistry* from 1903.

Contrary to other chemical conceptions of the ether, Mendeleev's was not a primary element from which the ordinary elements had evolved through condensation processes. It was just one element among others, sharing with them a position in the periodic table. Mendeleev argued that, for the ether to have real chemical significance, it must be perfectly permeable, ponderable but extremely light, corpuscular, and chemically inactive. He consequently placed it below helium in the new group of zero-valence elements. "The recognition of the ether as a gas," he wrote, "signifies that it belongs to the category of the ordinary physical states of matter. … All mystical, spiritual ideas about ether disappear" (Mendeleev, 1904, p. 14). From calculations based on astronomical evidence and the kinetic theory of gases, he suggested that the atomic weight of the ether was of the order $A = 10^{-6}$ or possibly less.

Although ponderable, since the ether element permeated everything it could not be weighed, and Mendeleev doubted if it ever would be possible to isolate it. He may have been aware of an earlier but failed attempt to isolate the ether due to the American chemist and engineer Charles Brush, who announced his startling discovery in 1898. According to Brush, he had discovered what he called "etherion," a new element with symbol Et and identical to the world ether. Brush's announcement in *Science* had scarcely appeared before it turned out that the etherion gas was nothing but mundane water vapor. In addition to the etherial element, which Mendeleev suggested to call either "x" or "newtonium," he also found space in his new periodic table for another inert element, the hypothetical coronium gas. Like helium, the belief in coronium was based on an unidentified spectral line, in this case $\lambda = 5876$ Å (Section 2.2.3). He estimated coronium's atomic weight to be at most $A = 0.4$.

Mendeleev was aware of J. J. Thomson's electron model of the atom (Section 2.1.5), but he did not believe in either this model or in the subatomic electron with a mass of the order $A = 10^{-3}$. Referring to what he called the "metachemical" electron, he stated that "it is my desire to replace such vague ideas by a more real notion of the chemical nature of the ether" (Mendeleev, 1904, p. 17). Whereas the great Mendeleev would have nothing to do with Thomson's electron, other *fin de siècle* chemists embraced it enthusiastically. According to Crookes, it was the long-sought primary particle, the ultimate chemical element. This was also the opinion of the Swedish physicist and chemist Johannes Rydberg, who in versions of the periodic table dating from 1906 and 1913 included the electron as an element on par with other elements. Ramsay, the celebrated chemist and Nobel Prize laureate, entertained similar ideas. In a paper of 1908 he argued that "Electrons are atoms of the chemical element, electricity; they possess mass; they form compounds with other elements; they are known in the free state, that is, as molecules; ... The electron may be assigned the symbol 'E'" (Ramsay, 1908). Contrary to Rydberg, Ramsay wisely avoided to place the electron in the periodic system.

2.1.4. CHEMICAL DARWINISM

Charles Darwin's bombshell of 1859, *On the Origin of Species*, was exclusively concerned with the evolution by natural selection of living beings, that is, animals and plants. Nonetheless, the controversial and much discussed theory also made a significant impact on inorganic sciences such as physics, chemistry, geology, and astronomy. Although "inorganic Darwinism" is, strictly speaking, an oxymoron, in the late nineteenth century many chemists and physicists adapted elements of Darwin's theory either implicitly or explicitly. What they adapted to their own sciences was typically evolution in a general sense and only rarely the idea of natural selection, which after all was the key point of Darwin's evolution theory. Scientists outside biology sometimes made use of phrases relating to natural selection, but then in a characteristic metaphorical and rhetorical manner.

As early as 1871, the British–Australian physicist Morris Pell, a professor at the University of Sydney, imagined an initial cosmic chaos out of which order grew by natural selection. "In the warfare among the molecules every enemy conquered would become the ally of the conqueror," he stated (Pell, 1872). "The molecules distinguished by numbers and strength of constitution would gradually gain the ascendancy by the destruction of weaker kinds." Assuming that the cosmic evolution depended on the characteristic wavelengths of the molecules, he reflected: "The possibility of cosmos evolving out of chaos (that is, the possibility that the material universe should become fitted to be the abode of organic life) may have depended upon whether or not a few constants were so arranged in the beginning as to satisfy a simple mathematical condition." It may not be too far-fetched to see in Pell's speculations an anticipation of modern ideas concerning the anthropic principle and the fine-tuned universe (Section 3.3.3).

Leopold Pfaundler, a respected professor of chemistry at the University of Innsbruck and a specialist in physical chemistry, concurred with Pell and developed the alleged Darwinian perspective in greater and more concrete detail. In studies of dissociation phenomena and chemical equilibria, he claimed to have found a profound analogy between the behavior of molecules and the mechanism of natural selection in the organic world. The Darwinian viewpoint helped him to understand why ammonium chloride dissociated by heating according to $NH_4Cl \rightarrow NH_3 + HCl$, whereas other compounds did not undergo a similar thermal dissociation. This and other processes he interpreted in agreement with the doctrine of survival of the fittest in the struggle for existence.

To return once again to Crookes' address of 1886, it was filled with Darwinian and quasi-Darwinian references. Although Crookes stressed throughout the analogy between chemical elements and living organisms, he also cautioned that the analogy was incomplete, for other reasons because "there cannot occur in the elements a difference corresponding to the difference between living and fossil organic forms." Writing a decade before the discovery of radioactivity, he confidently asserted that existing elements, although they had once been formed, could never become extinct like animals and plants of the distant past.

As far as the known elements were concerned — and at the time, there were about 70 of them — Crookes stated that they were the gradual outcome of an evolutionary process that might appropriately be described as a primordial struggle for existence: "Bodies not in harmony with the present general conditions have disappeared, or perhaps have never existed. Others — the asteroids among the elements — have come into being, and have survived, but only on a limited scale; whilst a third class is abundant because surrounding conditions have been favourable to their formation and preservation." In another passage of his 1886 address, Crookes noted that while some elements are abundant, others are very rare, a feature he also found a parallel to in the animal kingdom. He likened the rare-earth metals to "the Monotremata of Australia and New Guinea," a reference to the unusual egg-laying mammals platypus and echidnas.

The Danish chemist Julius Thomsen, professor at the University of Copenhagen and internationally recognized for his fundamental work in thermochemistry, shared with Crookes his fascination of the analogy between chemistry and evolutionary biology. In 1887, Thomsen indicated a possible explanation of the periodic system in terms of symmetries exhibited by hypothetical constituents of the supposedly composite atom. The symmetric patterns of the constituents expressed the "fitness" of the element in question. In this way, he thought he could explain, at least in principle, why elements only occur with certain atomic weights while others are missing, perhaps even prohibited. Why, for example, do the elements in the first period have $A = 7, 9, 11, 12, 16$, and 19, while there are none with $A = 8, 10, 13$, and so on?

Thomsen argued that when it came to the question of the origins of the elements, the chemist was in a position similar to the one of the biologist with regard to the constancy of the species. As Darwin and other naturalists had arrived at the evolution theory without being able to conduct experiments on the evolution of species, so the chemist had to rely on indirect arguments in the case of inorganic evolution:

> "Just as the biologist supposes that the right of the fittest has manifested itself in the evolution of species … the chemist has shown that the atomic weights of the elements do not form a successive series of numbers. Many numbers are missing among the known elements, and he is tempted to seek the answer in the right of the fittest which has manifested itself and only allowed the formation of atoms of a structure firm enough for a continuous existence. … The biologist believes that evolution from one species to another occurs through a series of generations … and the chemist must presumably adopt a similar hypothesis if he is to suggest the mechanism by means of which the transformation or development of an element leads to another element" (Kragh, 2016, p. 279).

Since Thomsen only published his considerations on Darwinian-like chemistry in Danish, they were unknown to the large majority of chemists.

Charles Darwin was vaguely aware that some chemists and physicists found his theory to be of relevance also in the domain of inorganic nature, but he did not share their belief and stuck to the theory's proper domain, the living world. On the other hand, one of his sons, the eminent geophysicist and astronomer George Howard Darwin, was convinced that evolution was an important common theme in all sciences. At the 1905 meeting of the British Association in South Africa, he dealt at length with the significance of his father's theory for the inorganic sciences. As he pointed out, "The origin and history of the chemical elements and of stellar systems now occupy a far larger space in the scientific mind than was formerly the case" (Darwin, 1905). At the time, the impact of the evolutionary perspective had greatly increased with the recent discoveries of radioactive decay and the electron as a constituent of all matter. What Crookes had speculated about in his address to the British Association nearly twenty years earlier had now, according to George Darwin, become a credible hypothesis approaching truth: "We are thus led to conjecture that the several chemical elements represent those different kinds of communities of corpuscles [electrons] which have proved to be successful in the struggle for life. If this is so, it is almost impossible to believe that the successful species have existed for all time."

2.1.5. ELECTRONS AND CHEMICAL EVOLUTION

As realized by G. H. Darwin and many other scientists (Mendeleev not being among them), by the early years of the twentieth century the atom had become a very different thing from what it was a decade earlier. Apart from radioactivity, the subatomic electron was chiefly responsible for the transformation. It is not generally appreciated that Joseph John Thomson's celebrated discovery in 1897 of the electron relied significantly on chemical reasoning in the Proutean tradition. Thomson identified cathode rays with a stream of electrons — or "corpuscles," as he called them — but instead of considering the particles to be liberated from the cathode metal he thought that they resulted from dissociations of the gas molecules in the intense electric field near the cathode. Thus, a remnant gas in the cathode-ray tube was necessary. Given that Thomson's experimental success depended on his ability to secure a very low pressure, it is ironic that, according to him, in a perfectly evacuated tube there would be no cathode rays at all! Thomson initially conceived the electrons to be a kind of chemical element and briefly contemplated the possibility of collecting the electrons and subjecting them to chemical analysis.

In a lecture of 30 April 1897, before he had succeeded in deflecting cathode rays electrostatically, Thomson explicitly associated his ongoing work with Prout's venerable hypothesis:

> "The assumption of a state of matter more finely divided than the atom of an element is a somewhat startling one; but a hypothesis that would involve somewhat similar consequences ... have been put forward from time to time by various chemists. Thus, Prout believed that the atoms of all the elements were built up of atoms of hydrogen, and Mr. Norman Lockyer has advanced weighty arguments, founded on spectroscopic consideration, in favour of the composite nature of the elements" (Thomson, 1897).

A few months later, he stated triumphantly that now he had found what previous workers had only suggested speculatively, namely Prout's long-sought universal constituent of matter. In his original experiments, Thomson only knew the charge-to-mass ratio e/m of the cathode ray electrons and therefore had to assume that the particles were subatomic, with a charge equal to that of the hydrogen ion H^+. Two years later, the assumption was confirmed when he succeeded in determining the charge of the electron, which led to a mass of the order one-thousandth of a hydrogen atom. From this, he concluded that atoms were aggregations of a large number of electrons placed in a massless and frictionless positive fluid.

In his books *Electricity and Matter* (1904) and *The Corpuscular Theory of Matter* (1907), Thomson used his atomic model to discuss why the hydrogen atom is the lightest known atom and why there is only a limited number of elements. In accordance with the earlier views of Crookes and other chemists, he illustrated the questions by referring to the inorganic evolution that had supposedly formed the elements during the long cosmic history. There was in Thomson's theory a definite cause for the nonexistence of certain potential atoms, namely that they were unstable and thus ruled out for physical reasons. The merely potential electron structures represented impossible and hence unrealized atoms. Thomson's idea was basically the same as the one suggested by J. Thomsen in 1887, which Thomson was unaware of.

Incorporating the evolutionary aspect Thomson argued that the matter existing in the far past differed completely from present matter, and likewise that in the distant future new kinds of matter would dominate the world. Thomson also discussed a possible explanation of the periodic system on the basis of his electron atom with the many electrons revolving in a system of concentric rings within the sphere of positive electricity. For example, for 22, 38, and 57 electrons, he found the equilibrium configurations 22(2, 8, 12), 38(2, 8, 12, 16), and 57(2, 8, 12, 16, 19), all of them with 2 electrons in the innermost ring and 8 and 12 electrons in the second and third ring, respectively. In Thomson's interpretation, this meant that the three model atoms had similar properties and hence were analogous to elements in the same group of the periodic system. Contrary to the later explanation, he associated the similarity with common internal groups of electrons (Section 1.2.5).

The British tradition in speculative atomic chemistry associated with scientists such as Crookes, Lockyer, and Thomson continued well into the twentieth century. An interesting but little known example is the theoretical physicist John Nicholson, who in works from the early 1910s developed an elaborate atomic model including elements from quantum theory. His model or theory assumed an atomic nucleus but was otherwise closer to Thomson's ideas of the structure of the atom. Of interest in the present context is Nicholson's conviction that terrestrial matter had evolved from simpler forms, what he called proto-atoms, that still existed in the stars but only there. They consequently could be studied only by means of the spectroscope. Nicholson operated with four proto-elements of which the simplest was "coronium" with $A = 0.51$ and consisting of a nuclear charge $+2e$ surrounded by a ring of 2 electrons. His "nebulium" atom had nuclear charge $+4e$ and $A = 1.63$.

Considerations of an astrophysical and astrochemical nature were of crucial importance to Nicholson's model, which for a brief period of time attracted attention as an alternative to Bohr's quantum atom (see Section 1.2.2). Foreshadowing much later developments in nuclear and particle physics, Nicholson argued that, to understand the architecture of atoms and fundamental physics generally, the physicist would have to look to the heavenly regions. In 1913, he characterized astrophysics as "an arbiter of the destinies of ultimate physical theories" (Nicholson, 1913). His chief argument was that "when an astrophysicist discovers hydrogen in a spectrum, he is dealing with hydrogen in a simpler and more primordial form than any known to a terrestrial observer." The general idea that astrophysical and cosmological studies reveal secrets about nature that cannot be revealed in laboratory studies turned out to be fruitful, but Nicholson's own work was unproductive and soon forgotten.

After World War I, astrophysics and astrochemistry changed considerably, becoming more professionalized and less speculative branches of science. Although chemistry remained overwhelmingly a laboratory science, a minority of chemists maintained an interest in the celestial regions. They conceived the stars to be nature's own huge laboratories, outside the reach of the chemist's traditional means but interesting for this very reason. Gilbert Newton Lewis, a distinguished and versatile American chemist, was a pioneer in atomic structure, the electron theory of the chemical bond, and chemical thermodynamics. He also had an interest in cosmic processes and the origin of the elements. In a paper from 1922 published in an astronomical journal, he suggested that not only had astronomy much to learn from chemistry, the latter science had also much to learn from astronomy. Lewis argued as follows:

> "While the laboratory affords means of investigating only a minute range of conditions under which chemical reactions occur, experiments of enormous significance are being carried out in the great laboratories of the stars. It is true, the chemist can synthesize the particular substances which he wishes to investigate and can expose them at will to the various agencies which are at his command; we cannot plan the processes occurring in the stars, but their variety is so great and our methods of investigation have become so refined that we are furnished an almost unbounded field of investigation" (Lewis, 1922).

Much of the chemically oriented work in astronomy during the interwar years was concerned with charting the abundances of the elements, a line of work that became very important in nuclear astrophysics and cosmology (Section 3.3.1). Generally, the chemical elements and not their compounds were the exclusive focus, because of which reason many mainstream chemists hesitated in recognizing astrochemistry as belonging to chemistry proper. It was long uncertain whether molecules existed in interstellar space, but between 1937 and 1941 a few diatomic molecules (CH, NH, CN, CH^+) were identified by means of their characteristic spectra. The first identification may have been the CH molecule or radical, which was detected in 1937 by two Belgian scientists, the astronomer Pol Swings and the physicist Léon Rosenfeld. After World War II, the number of interstellar molecules grew rapidly, and today more than 200 are known. Most of them are organic and some quite complex. Examples are acetic acid CH_3COOH, acetone $(CH_3)_2CO$, and the amino acid glycine CH_2NH_2COOH. Another and simpler molecule is the triatomic hydrogen ion H_3^+, the history of which is the subject of Chapter 2.3.

2.2

Auroral Chemistry

2.2

Auroral Chemistry

As indicated by its name, polar light (aurora polaris) is a natural light display predominantly seen in the regions around the Arctic and the Antarctic. Although the phenomenon is symmetric between the magnetic north and south, the northern light or aurora borealis is much better known and has been studied much longer than the southern light or aurora australis. The first scientific observations of the southern light date from the 1770s. Aurorae can on rare occasions be seen far from the high-latitude regions, such as illustrated by the earliest written accounts of them. The oldest known record of the aurora borealis is from the unlikely city of Babylon in the present-day Iraq and has been dated to the night between 12 March and 13 March 567 BC. Historians have identified the description of an aurora in a clay tablet dated "Year 37 of Nebuchadnezzar, King of Babylon." Aurorae were also known to the Greeks and are, for example, described in Aristotle's important work *Meteorologica* from around 330 BC. According to Aristotle: "Sometimes on a fine night we see a variety of appearances that form in the sky: 'chasms' for instance and 'trenches' and blood-red colors. These, too, have the same cause. For we have seen that the upper air condenses and sometimes takes on the appearance of a burning flame, sometimes that of moving torches and stars" (Aristotle, 1984, Vol. 1, p. 560).

Until the scientific revolution, aurorae belonged essentially, like comets and meteor showers, to myths and superstitious belief. They were often supposed to be omens of disaster. Only during the nineteenth century did auroral research emerge as a branch of science cultivated by a growing number of astronomers, physicists, meteorologists, chemists, and amateur scientists. At the turn of the century, it had become a major field of scientific study, albeit one lacking disciplinary unity and with very little institutional support. Many scientists studying the aurora came from meteorology and physics, and only few from chemistry. Indeed, the term "auroral chemistry" may seem strange, for how can the aurora by investigated by chemical means? Nonetheless, problems of a chemical nature were an important part of the development that led to an understanding of the remarkable colors displayed by aurorae.

The key problem was to understand the auroral spectrum in terms of the chemical elements responsible for it, and by definition elements belong to chemistry. More specifically, a green spectral line of wavelength ca. 5577 Å appeared in all aurorae, and the identification of this line, the holy grail of auroral science, turned out to be frustratingly difficult. It took more than half a century before the problem was solved, and during this long period a few scientists even proposed the existence of new elements specific to the aurora in the upper atmosphere. This chapter focuses on the search for the element responsible for the green aurora line.

2.2.1. AURORA BOREALIS

Although the northern light had been known since antiquity, the name aurora borealis first appeared in 1619, possibly coined by Galileo Galilei. It took another century before the phenomenon caught the attention of natural philosophers in the new scientific tradition. The first description of an aurora that can be reasonably called scientific was published in 1707 by the Danish astronomer Ole Rømer, who is best known for his discovery of the finite velocity of light thirty years earlier. Rømer had witnessed the aurora in Copenhagen, and he described it tersely and objectively, without trying to explain the nature and origin of the phenomenon. A decade later, another famous astronomer, Edmond Halley, observed a magnificent aurora in London. In his article in *Philosophical Transactions*, he speculated that the colorful display was the result of "magnetic effluvia" emanating from inside the Earth. The most comprehensive and influential of the works of this early period was the French physicist Jean Jacques de Mairan's *Traité Physique et Historiques de l'Aurore Boréale* (Physical and Historical Treatise on the Aurora Borealis) first published in 1732. De Mairan thought that the aurora was a manifestation of a hypothetical subtle medium penetrating all of space.

During the following century, knowledge of the aurora greatly improved, in part because of more systematic observations and in part because of progress in the sciences of electricity, magnetism, and optics. On the instrument side, spectroscopy entered studies of the aurora in the late 1860s and photography only some thirty years later. The first photograph of an auroral spectrum dates from 1898. It was particularly important when in was realized about 1850 that the recently discovered 11-year sunspot cycle was matched by an almost identical cycle in geomagnetic storms and also in auroral activity. There could be little doubt that the sunspots, or the Sun generally, were the cause of the aurorae. Moreover, by the 1880s it became accepted that aurorae had an electrical origin and were possibly the result of atmospheric atoms and molecules stimulated by collisions with electrical particles emitted by the Sun. It was still unknown where in the atmosphere the colors of the aurorae were formed, but at around 1908 the Norwegian mathematician and aurora expert Carl Størmer showed that it took place at a height varying between approximately 100 km and 300 km, in what presently is called the thermosphere.

Around the turn of the century, the generally favored view was that solar cathode rays, since about 1900 known as electrons, were the principal agents of the aurorae. The leading advocates of this view, which they expounded in different and rival versions, were the Norwegian physicist Kristian Birkeland and the Swedish physical chemist Svante Arrhenius.

Birkeland's experiments with electrical discharges and the action of strong magnets on cathode rays led him in 1896 to suggest that "the [cathode] rays come from cosmic space and are in particular absorbed by the Earth's magnetic pole, and that in some way or other they must be attributed the Sun" (Birkeland, 1896, p. 512).

According to Birkeland, aurorae and magnetic storms were caused by high-energy charged particles from the Sun being deflected and accelerated in the magnetic field of the Earth. When the particles reached the upper strata of the atmosphere, they would be slowed down, in the process exciting and ionizing the atoms, which as a result would emit light of the colors characteristic of the aurora. Birkeland likened the electrons and alpha particles emitted by the Sun to a huge radioactive disintegration, except that the energies of the solar rays were much higher than known from radioactive decay. To support his hypothesis, Birkeland conducted a series of spectacular experiments consisting of a magnetized steel sphere, a so-called terrella (little Earth), placed in a large vacuum chamber. The steel sphere simulated the Earth and the vacuum or low-pressure air around it represented the upper atmosphere enveloping the Earth. By bombarding the sphere with cathode rays, Birkeland succeeded in creating artificial aurorae strikingly similar to those observed in nature. The experiments attracted much public attention, whereas many scientists were less impressed. They considered Birkeland's theories to be speculative and his claims of laboratory support to be nothing but unconvincing analogies.

By 1900, Arrhenius had largely left physical chemistry and instead changed his research area to what at the time was called "cosmical physics," a thriving interdisciplinary field mainly comprising meteorology, astronomy, and physics, but also with contributions from geology, oceanography, and chemistry. The aurora played an important role in cosmical physics. Arrhenius hypothesized that aurorae were due to very fine, electrically charged dust particles expelled from the Sun and carried towards the Earth by the strong solar radiation pressure. For spherical dust particles of density 1 g/cm^3, he estimated that they must have a diameter of less than 10^{-4} cm to escape the Sun's gravitational attraction. When the electrified particles sailing on the radiation pressure reached the Earth's upper atmosphere, they would be discharged and in the process cause the atmospheric atoms to emit light in the colors characteristic of aurorae.

Although cathode rays played an important role in Arrhenius' theory, it was indirect and quite different from the one in Birkeland's theory, where the electrons traveled from the Sun to the Earth. According to Arrhenius' alternative, the dust particles from the

Sun would charge the atmosphere with negative electricity, and as a result discharges followed by the formation of cathode rays would occur. Thus, the cathode ray electrons involved in the process were terrestrial and produced in the atmosphere; they did not originate in the Sun, as in Birkeland's theory. During about 1910, most experts adopted a theory of the origin of the aurora similar to either Birkeland's or Arrhenius'. Although these theories offered a mechanism for the generation of aurorae in the upper atmosphere, they had nothing to say about the colors appearing as spectral lines in the auroral spectrum. This question could only be addressed by spectroscopic studies, an area of research that was of limited relevance for the theories of the two Scandinavian scientists.

2.2.2. THE RIDDLE OF THE GREEN LINE

In 1859, spectrum analysis experienced a breakthrough as the result of a unique collaboration between two eminent scientists at the University of Heidelberg, the chemist Robert Wilhelm Bunsen and the physicist Gustav Robert Kirchhoff. With the new spectroscope, they showed that the emission spectra were reliable fingerprints of chemical elements. Not only could spectra be used to identify small amounts of known elements, they could also be used to predict new ones. Within a year, the chemical power of the spectroscope was dramatically demonstrated with the discovery of two new metallic elements, cesium and rubidium. Thallium followed shortly thereafter. Kirchhoff immediately pointed out that the new technique had important astronomical consequences. In a paper from 1860, he demonstrated that sodium was a constituent of the solar atmosphere and soon thereafter he also identified the presence of iron, magnesium, copper, zinc, nickel, and barium.

Despite their splendor, aurorae shine with low intensity and their spectra are not easily recorded. The first spectroscopic observation of an aurora was announced in 1868 by the Swedish physicist Anders Johan Ångström, who found a bright greenish line of wavelength 5567 in the unit named after him (1 Å = 1 angstrom = 0.1 nm). Ångström's observations gave rise to a great deal of activity and many measurements of the spectrum of the aurora borealis. The wavelength of the green line first reported by

Ångström was re-determined by numerous later scientists, eventually to stabilize at $\lambda = 5577$ Å. Interferometric measurements made in 1923 gave the more precise value $\lambda = 5577.350 \pm 0.001$ Å. The nature of the green line, often referred to as just the auroral line, remained a puzzle for several decades. The general line of attack was to make experiments with different gases in discharge tubes at varying temperature and pressure, and compare the spectra with those obtained from observations of aurorae. The method was thus basically the same that in 1860 had proved the existence of sodium in the Sun's atmosphere.

Since the early 1870s, many researchers were busy with studying the spectrum of the aurora. This kind of work resulted in an extension of the number of spectral lines and increasingly precise wavelengths, but not in a satisfactory understanding of the chemical nature of the substances responsible for the spectrum. It was often assumed that the green line was due to either oxygen or nitrogen atoms which, at the low pressure and temperature residing in the upper atmosphere, were excited by electrical actions coming from the Sun. However, despite many attempts, no one succeeded in reproducing the green line in the laboratory, and thus its nature remained an unsolved problem. The state-of-art in auroral spectroscopy in the early twentieth century is reflected in a detailed article on the aurora polaris in the 1911 edition of *Encyclopedia Britannica*. According to this article, the most complete record of the spectrum was obtained by a team of Swedish scientists on Spitsbergen who found no less than 158 auroral lines, many of which coincided with oxygen and nitrogen lines. However, this was not the case with the green line. In a mood of despair, the German physicist Heinrich Kayser, possibly the world's leading expert in spectroscopy, concluded that "We know nothing at all about the chemical origin of the lines of the polar light" (Kayser, 1910, p. 58).

For a period of time, nitrogen was the favorite source of the green line, but closer inspection showed that it was impossible to obtain a precise match between the auroral line and nitrogen lines produced in the laboratory by means of electrical discharges in dilute gases. Among the alternatives more or less seriously discussed were that the line might be due to a fluorescent form of argon or that it had its origin in metallic dust particles arising from the combustion of meteorites in the upper atmosphere. None of these hypotheses were vindicated, and also the popular krypton hypothesis proved to be a blind alley.

The inert gas krypton had been discovered in 1898 by William Ramsay and Morris Travers, and for a while it appeared to be the best candidate for several of the yellow and green lines in the aurora spectrum. The wavelength of an intense line in the krypton spectrum was measured to 5570.4 Å, which many scientists found to be sufficiently close to the green line to make the identification convincing (at the time the best value for the green aurora line was 5571.0 Å). In 1907, an English meteorologist concluded optimistically that "There seems now little doubt that the chief line in the aurora, i.e., Ångström's green line, must be assigned to krypton, and there is considerable probability that the red line is due to neon" (Watts, 1907). However, the optimism was premature as well as problematic. For one thing, krypton with atomic weight 83.8 was known to be a very rare constituent of air and it was difficult to imagine why such a relatively heavy gas should be abundant in the upper atmosphere; for another thing, not all of krypton's bright lines could be found in the spectrum of the aurora.

Ramsay, the Nobel Prize-rewarded discoverer of helium and co-discoverer of krypton, cautiously suggested that the question was still unsettled. In an essay from 1909, he summarized:

> "I am not able to decide yet whether the [aurora] lines are all due to krypton or to ... nitrogen. Certainly there is a striking similarity between the nitrogen spectrum and that of the aurora; and, on the other hand, the lines of krypton, though sufficiently coincident with those of the aurora to satisfy criticism, leave other bright lines of the krypton spectrum unaccounted for. ... Experiments on this matter are not yet decisive" (Ramsay, 1909, p. 213).

There was yet another possibility, although perhaps a far-fetched one, namely to assume that the green line and possibly also other of the unidentified auroral lines were due to an element unknown to the chemists. And indeed, a couple of proposals of auroral elements were actually made, one as early as 1869 and the other as late as 1911. From a modern point of view, such proposals may appear suspect, to say the least, but before 1915 they were legitimate speculations not ruled out by atomic theory or the periodic system. Mendeleev's original system did not preclude the existence of new elements that existed only in the heavenly regions. Hypothetical elements such as coronium and nebulium, suggested on the basis of unidentified spectral lines, held considerably credibility, and in 1895 helium, until then just another hypothesis, was discovered in terrestrial sources and thereby turned into a real element. If helium were accepted as real, why not an auroral element? In his presidential address of 1902 to the British Association for the Advancement of Science, the esteemed chemist James Dewar suggested that some of the lines in the aurora spectrum might perhaps be due to a gaseous element yet to be discovered in the atmosphere. Why not?

2.2.3. AURORAL ELEMENTS?

The first suggestion of a distinct auroral element, occasionally referred to as "aurorium," came from an unlikely source, namely the later so famous American philosopher and logician Charles Sanders Peirce (see also Section 3.1.4 for his ideas of curved space). In 1869, 30-year old Peirce worked with spectroscopic and photometric studies at Harvard College Observatory, and on 15 April the same year he measured the spectrum of an aurora. In an anonymous review from the same year of a textbook on spectrum analysis, he suggested that the green line might be due to a very light gas with atomic weight 0.07, about one-fourteenth of that of hydrogen. "This substance must extend down to the surface of the earth," Peirce wrote (Kragh, 2010). "Why, then, have chemists not discovered it?" His answer was that the substance "must be a very light elastic gas ... [and] be as much lighter than hydrogen as hydrogen is than air." Peirce's bold hypothesis attracted almost no attention, appearing as it did in an unsigned review in *The Nation*, an American weekly magazine unknown to European scientists. When the suggestion of a subhydrogenic auroral element was revived some forty years later, it was in ignorance of Peirce's earlier speculation.

Alfred Wegener is possibly the most famous earth scientist of the twentieth century, his fame resting on the revolutionary theory of continental drift which he outlined in 1912 and three years later presented in full in his classic work *Die Entstehung der Kontinente und Ozeane* (The Origin of Continents and Oceans). Trained in astronomy and meteorology — but not in geology or geophysics — Wegener had a strong interest in atmospheric science, which was his main field of study just before he turned to the drifting continents. In 1911, he published a major work in the field, a wide-ranging monograph with the title *Thermodynamik der Atmosphäre* (Thermodynamics of the Atmosphere) in which he investigated, by means of physical and chemical theories, the structure and material content of the atmospheric layers.

Measurements showed that the intensity of the $\lambda = 5570$ Å green line increased with the height of the aurora and completely dominated the so-called steady arcs at high altitude. To Wegener, this was evidence that the line originated in a light gas only found in the uppermost regions of the atmosphere, such that he first argued in a paper of 1910. The following year, he suggested that the hypothetical gas was a new chemical element similar to the "coronium" of which the solar corona was supposedly made. The hypothetical coronium was only known by the unidentified green coronal line of $\lambda = 5203$ Å, just as helium was only known by the yellow line of $\lambda = 5876$ Å before 1895. The corona line was first detected by the American astronomer Charles Young in 1869 and still around 1910 coronium was widely accepted by astronomers. There were even a few claims of having detected traces of the element in volcanic gases and other terrestrial sources.

Wegener was in part inspired by an unusual revision of the periodic system that Mendeleev had proposed in 1902 and that included two gaseous elements lighter than hydrogen. One of the new elements might be coronium with atomic weight 0.4 and the other might be the much lighter world ether of which Mendeleev was a staunch advocate and to which he assigned an atomic weight less than one-millionth (Section 2.1.3). On his part, Mendeleev received inspiration from Dewar's 1902 suggestion of a possible auroral element. While the large majority of chemists and physicists chose to ignore Mendeleev's unorthodox chemical ether speculations, Wegener took them seriously as far as coronium was concerned. He did not comment on Mendeleev's ethereal element.

In publications from 1911, Wegener introduced what he called "geocoronium" as the source of the aurora spectrum. He offered a new picture of the chemical composition of the atmosphere, which for very high altitudes differed drastically from the usual one. According to Wegener, the low-altitude nitrogen–oxygen sphere was followed by a hydrogen sphere which at heights greater than ca. 150 km was increasingly mixed with the hypothetical geocoronium and also contained small amounts of helium. At a height of 200 km, the atmosphere would consist of equal amounts of hydrogen and geocoronium, whereas at a height of 500 km the distribution would be 93% geocoronium and 7% hydrogen.

In a paper of 1911, Wegener, referring to the green aurora line, expressed his belief that "in this way it is possible, for the first time, to establish some order in the confusing chaos of contradictory observations and opinions" (Wegener, 1912). Following Mendeleev, he assumed geocoronium to be monatomic and with atomic weight 0.4. Moreover, he estimated that at a height of 200 km geocoronium's partial pressure was equal to that of hydrogen, and from this he calculated that the atmosphere at sea level should include 0.00058% geocoronium by volume. This was a small amount indeed, but not very much smaller than the amount of hydrogen, which he cited as 0.0033%. He optimistically stated that perhaps geocoronium might be detected by chemical means. As to the nature of the new gaseous envelope of the Earth, Wegener thought it was similar to or perhaps identical with the solar coronium. Although the bright 5570 aurora line did not coincide with the 5203 coronium line, he argued that the difference might be ascribed to the different excitation mechanisms in the solar and terrestrial coronas. He therefore concluded that the two gases were most likely identical.

Published in a monograph and in two of the leading journals of physics and chemistry, Wegener's geocoronium hypothesis was noticed by contemporary scientists. However, it was coolly received, one of the few exceptions being Birkeland, who found it to be interesting but did not endorse it. Most chemists were unwilling to consider new element of the kind proposed by Mendeleev and Wegener, and meteorologists and other aurora specialists thought that the green line could be explained without the drastic assumption of a new gas enveloping the Earth. After all, there were no independent evidence of either coronium or geocoronium, and so the hypothesis seemed to be ad hoc and rest on a circular argument. To justify one hypothetical element by means of another hypothetical element carried no conviction. Wegener briefly referred to the high-altitude geocoronium atmosphere in his first and today famous paper on continental drift of 1912, but after that time he no longer defended the hypothesis. Geocoronium vanished from the scene of science.

It took longer for coronium to follow the fate of geocoronium, as it only vanished for good in 1939. In that year, Walter Grotrian in Germany and Bengt Edlén in Sweden proved that the corona emission line of 5303 Å was due to a rare electron transition in Fe^{13+} ions where half of the electrons in the neutral iron atom are missing. Similarly, the previously unidentified red corona line of 6375 Å was explained in terms of Fe^{9+} ions.

2.2.4. A FAILED RESEARCH PROGRAM

The Norwegian physicist Lars Vegard, a former assistant of Birkeland and since 1918 his successor as professor at the University of Christiania (Oslo), was generally recognized as the world's leading authority in auroral spectroscopy. Apart from his important work on the aurora, he also contributed to the quantum theory of atomic structure by developing in 1918 electron configurations for all the elements in the periodic system. Vegard was, from an early date, convinced that the upper atmosphere consisted mainly of nitrogen, and that earlier hypotheses of helium, hydrogen, krypton, or unknown gases (such as geocoronium) were wrong. Since the early 1920s, he focused on a systematic research program with the aim of solving the green line mystery and proving that the line was caused by electrically excited nitrogen atoms. Vegard originally studied the auroral spectrum *in natura,* using advanced spectrographs in northern Norway. He measured 35 distinct lines, most of which could be assigned to nitrogen, and for the still unexplained green line he obtained a best value of $\lambda = 5577.6$ Å.

Vegard realized that the atoms in the very cold and rarefied upper atmosphere must exist in an unusual state that could not be examined in ordinary laboratory experiments. In about 1923, he proposed a theory of the higher strata of the atmosphere and also of the unusual state of the nitrogen atoms which he supposed populated these strata. The theory was quite different from the one commonly accepted by geophysicists and meteorologists. According to Vegard, the Earth was enveloped in an outer dust atmosphere, an electrified layer of tiny frozen dust particles starting at a height at about 90 km. Arguing that his favorite hypothesis agreed with all known astronomical and meteorological phenomena, he was particularly intrigued by its connection to the spectrum of the aurora. If he could reproduce the lines of the aurora, and in particular the controversial 5577 line, he would have killed two birds with one stone: he would have solved the old problem of the chemical composition of the aurora and also have confirmed the new picture of an electrified upper dust atmosphere.

To test the hypothesis, Vegard went to Leiden in the Netherlands to do experiments at Heike Kamerlingh Onnes' famous low-temperature laboratory, which at the time was the most advanced laboratory of its kind. It was here that helium was first liquefied and superconductivity discovered, in 1908 and 1911, respectively, and it was for this work that Onnes was awarded the 1913 Nobel Prize. Vegard made elaborate experiments in Leiden with thin layers of frozen nitrogen exposed to high-voltage cathode rays of an energy up to 10,000 electron volts. Undoubtedly to his great joy, he noticed a brilliant greenish light looking suspiciously like the one known from the aurora.

Vegard's work to reproduce the conditions of the auroral light was an example of so-called mimetic experiments in which natural phenomena are scaled down and reproduced in the laboratory. This kind of experimental approach dates from the eighteenth century when several scientists made experiments to create "artificial aurorae" by means of electrical discharges or luminescent substances. During the next century, mimetic experiments became common in astrophysics, geophysics, and related areas. Thus, ever since the 1870s there had been a well-established tradition within solar and stellar spectroscopy to obtain knowledge of stellar atmospheres by reproducing in discharge tubes the spectral lines observed from the stars. The same approach had been adopted by researchers attempting to understand the composition of the aurora. Birkeland's terrella experiments, well known to Vegard, were a beautiful example of mimetic science. The experiments in Leiden were in some respects a much refined version of the mimetic experiments that he and his former professor had performed in Oslo before World War I.

What Vegard found was actually a band spectrum with three maxima very close to the green line and not the line itself. Nonetheless, convinced as he was of his hypothesis he argued that he had found a limit in the band spectrum of 5578.6 Å and that the difference of about 2 Å could be explained as an artifact, a disturbing effect in the experiment. Within half a year, he felt sure that now he had solved the riddle of the green line. As he triumphantly reported to *Nature*, "the typical auroral spectrum is emitted from solid nitrogen, and thus my hypothesis with regard to the constitution of the upper atmosphere has been confirmed" (Vegard, 1924). Moreover, his experiments made with liquid hydrogen and helium indicated that the aurora-like spectrum only appeared at temperatures below 35 K. From this, Vegard concluded that the auroral region in the atmosphere at about a height of 100 km was correspondingly cold. At the same time, British physicists and meteorologists favored a temperature of approximately 300 K!

Vegard's widely announced discovery claim was received with interest, but also with caution. Physicists in Germany and England criticized the work of their Norwegian colleague, which they found to be unconvincing and contrary to their own views of the upper atmosphere and preference for oxygen rather than nitrogen. In 1924, the Leiden laboratory was no longer the only laboratory able to produce liquid hydrogen and helium. The previous year a new facility had been established in Toronto under the leadership of John C. McLennan, an expert in radioactivity and spectroscopy who also had an interest in the aurora.

Together with his research student Gordon Shrum, in 1924 McLennan conducted a series of experiments of essentially the same kind as those Vegard made in Leiden. Shrum recalled: "We had first-rate spectroscopy equipment; we could generate the low temperatures — we could reproduce anything in the upper sky. We could produce all the other lines in the aurora, but not this green line" (Shrum, 1986, p. 43). Investigating the spectra of various gases exposed to cathode rays at temperature about 20 K, the two Canadians found several lines in nitrogen, but none of them came even close to the 5577 aurora line. In a paper from early 1925, they consequently reported that they had refuted Vegard's hypothesis. Vegard, on the other hand, denied that this was the case and suggested that the experiments in Toronto were incomplete and less reliable than his own. While McLennan and Shrum did not come up with an alternative to the nitrogen dust hypothesis in their early experiments, this is what they did in later experiments of 1925, which not only challenged Vegard's discovery claim, but proved it wrong.

2.2.5. THE GREEN LINE REVEALED AND EXPLAINED

The laboratory discovery of the green auroral line that occurred in March 1925 was serendipitous. McLennan and Shrum believed, as did many meteorologists and geophysicists, that the upper atmosphere consisted predominantly of helium with only small amounts of oxygen and nitrogen. For this reason, their working hypothesis was that the green line originated in some unusual state of helium unknown to laboratory physics. The early experiments in Toronto were done by Shrum, who in agreement with the helium assumption investigated the spectrum of helium at very low pressure. After many failed attempts to obtain the green line, in an experiment of late February 1925 he finally succeeded — except that the line mysteriously disappeared in subsequent repetitions of the experiment! After two weeks of increasing frustration, the line did turn up again, according to Shrum's recollection in this way:

> "I thought I was going to have a nervous breakdown. At last I was so desperate that I tried a most unscientific approach. I decided that I would get up at the same hour of the morning as I had on the great day, have the same breakfast, go over to the university at the same hour, put on the equipment in the same way, and look in the tube at exactly the same time. I did all that, looked in, and the line was there" (Shrum, 1986, p. 46).

What had happened was that when Shrum first saw the line he had not purified the gas properly and therefore unknowingly worked with helium contaminated with oxygen and nitrogen. Realizing that lack of purity was essential, he changed to helium–oxygen and helium–nitrogen mixtures with the result that the green line turned up in the first mixture but not in the second. Nor did it turn up in pure helium. This work, and therefore the original reproduction of the green line, was due to Shrum alone, whereas McLennan only entered the experiments subsequently. Having confirmed the observation of his research assistant, McLennan quickly announced the discovery in a letter to *Nature,* but without adding Shrum as a coauthor. The full report followed some months later in a detailed article in *Proceedings of the Royal Society,* this time with both McLennan and Shrum as authors.

The two discoverers pointed out that the green line did not come from helium but from the oxygen component, and they reported a wavelength of the line in full agreement with the best determination of the auroral line found in the night sky. This determination obtained by means of interferometry was $\lambda = 5577.350 \pm 0.001$ Å, whereas the McLennan–Shrum laboratory result was 5577.35 ± 0.15 Å. On the basis of their experiments, McLennan and Shrum further suggested that the composition of the atmosphere at height 110 km was 95% helium and 5% oxygen. Still in their 1925 discovery paper, they concluded that the green line only turned up in oxygen when mixed with helium, but later experiments in Toronto proved that the role of helium was merely to enhance the intensity of the line, which was observed even in pure oxygen. In 1927, McLennan narrowed down the wavelength of the green oxygen line to $\lambda = 5577.341 \pm 0.004$ Å, with an optical line width of $\Delta\lambda - 0.030$ Å.

The reproduction and subsequent theoretical explanation of the green line was a breakthrough in auroral physics and the high point in McLennan's career. In a letter to McLennan, the Indian physicist and Nobel laureate of 1930, Chandrasekhar Raman, expressed his admiration in a somewhat exaggerated way: "Your work on the green line of the Aurora is one of the most sensational achievements in modern physics" (Eve, 1935). In 1927 McLennan received the prestigious gold medal from the Royal Society and shortly before his death in October 1935, he was even nominated for a Nobel Prize. The same year Vegard, his rival in auroral spectroscopy, was nominated, although in his case for his contributions to the physics of the aurora in general. The prize of 1935 went to James Chadwick for the discovery of the neutron. In Oslo, Vegard was unwilling to admit that the Canadian physicists had really solved the riddle of the green line and thereby made his own and earlier alternative obsolete. For a couple of years, he maintained his claim that the solid nitrogen hypothesis was the true solution and that the results obtained in Toronto did not prove the green line to have its origin in oxygen atoms. Only in about 1933 did he grudgingly accept defeat.

Whereas McLennan harvested the fruits for the discovery of the green line, his junior partner Shrum was soon forgotten. And yet the key discovery in the early spring of 1925 was mainly due to the 29-year-old postdoc Shrum, whereas McLennan only played a secondary role. Shrum left Toronto for the University of British Columbia in August 1925, in part as a result of what he felt was McLennan's unfair appropriation of the discovery. The discovery of the green line, principally made by Shrum but generally credited to McLennan, is a clear-cut example of what in history and sociology of science is known as the "Matthew effect." This effect refers to the inordinate amount of credit that is often assigned to highly reputed scientists at the expense of their lesser known collaborators or students. Or to quote Matthew 25: 29, "To all those who have, more will be given, and they will have an abundance; but from those who have nothing, even what they have will be taken away."

Up to 1926, all work done on the spectrum of the northern lights, including that of Vegard and McLennan, had been experimental and observational. However, a complete understanding of the green line would not only require laboratory reproduction but also theoretical justification. Fortunately, just at that time the new quantum mechanics arrived on the physics scene offering a much improved understanding of atomic spectra.

Only certain transitions between the energy levels of an atom are allowed, while others are "forbidden" because they violate one or more selection rules. Nonetheless, it was known since the early 1920s, in connection with studies of the helium spectrum, that such forbidden transitions sometimes occur with very low intensity, which was explained by introducing relatively long-lived, so-called metastable states in addition to the ordinary energy states with a lifetime of about 10^{-8} seconds. Forbidden transitions are not, strictly speaking, forbidden, but simply so improbable that they will not occur under normal laboratory conditions, where frequent collisions between atoms will prevent the formation of metastable states. By applying the quantum-mechanical theory of spectra to all possible transitions in an oxygen atom, McLennan and his collaborators in Toronto suggested that the green auroral line was due to a transition between two metastable quantum states. The suggestion received support from other physicists with the result that, latest by 1930, there existed a detailed explanation of the green line in terms of atomic theory.

The green line was not the only puzzling spectral line from the heavens that was explained as due to forbidden transitions between metastable states. In 1927, the American astrophysicist Ira Bowen explained the nebulium lines 5997 Å and 4959 Å along similar lines of reasoning, in this case as due to forbidden transitions in the oxygen ion O^{2+}. The existence of a hypothetical "nebulium" element was proposed in the late nineteenth century, and for a couple of decades it was taken seriously by a minority of astronomers and physicists. As mentioned, the coronium lines 5203 Å and 6375 Å were explained only in 1939, and then as the result of forbidden quantum transitions in iron ions. Although the quantum theory of complex atoms played a decisive role in all three problems, in the case of the green line it only entered *post factum*, after the line had been reproduced experimentally. It worked as a theoretical justification of what was already known from laboratory experiments compared to measurements of the auroral spectrum. By the late 1920s, the enigmatic green line of the aurora borealis was no longer an enigma. But the solution did not come easily, for it took nearly sixty years until a full understanding of the line was obtained.

2.3

Triatomic Hydrogen

2.3

Triatomic Hydrogen

Normal molecular hydrogen is diatomic, consisting of H_2 molecules. Students of chemistry are also familiar with the highly reactive monatomic hydrogen, which in a historical context sometimes appear as nascent hydrogen or hydrogen *in statu nascendi*, but they may never have encountered the triatomic molecular form H_3 or its corresponding ion H_3^+. These unusual forms of hydrogen were originally suggested by J. J. Thomson in the early 1910s and from about 1920 to 1935 they were eagerly studied by chemists as well as physicists. The interdisciplinary research field resulted in approximately one hundred scientific papers of which half were predominantly of a chemical nature and the other half more focused on physics. Many of them appeared in journals of physical chemistry or, since the early 1930s, chemical physics. By far, most of them were written by American, British, and German authors. Among those who contributed to the field were several Nobel Prize laureates or future laureates, including notables such as Thomson, N. Bohr, J. Stark, F. W. Aston, H. C. Urey, L. Pauling, and G. Herzberg.

Measurements with the new mass spectrometer quickly confirmed Thomson's hypothesis of the H_3^+ ion, whereas it proved much more difficult to establish the existence of the neutral molecule. During the 1920s, many chemists investigated so-called active hydrogen, widely believed to be H_3, and for a time triatomic hydrogen seemed to be on the verge of discovery. However, with critical reexaminations and improved experiments belief in H_3 declined, and by 1935 the hypothesis was practically dead. Nonetheless, many years later the "dediscovery" was followed by a rediscovery, namely when Herzberg found convincing evidence in the form of infrared spectral lines originating from the molecule.

Triatomic hydrogen, whether in its ionic or neutral form, was for a long time thought to be of no relevance to astronomy. After all, why should it? Following the spectral identification of H_3^+ in laboratory experiments dating from 1980, the attitude changed drastically. While H_3 belongs to the artificial world of the laboratory, the ion turned out to be abundant in interstellar space and elsewhere in the universe. Today the study of H_3^+ is even more interdisciplinary than in the past and of great interest to astrophysicists and astrochemists in particular. Moreover, the little-known historical case of triatomic hydrogen is also instructive from a philosophical perspective, for other reasons because it illustrates the evidential nature of scientific knowledge. By weighing evidence for and against the hypothesis, scientists concluded that there was no good reason to assume that H_3 existed elsewhere than in the mind of the theorists. The conclusion turned out to be wrong, thus illustrating that consensual knowledge is not by itself a guarantee of truth.

2.3.1. EARLY THEORIES AND EXPERIMENTS

From about 1906 to 1914, J. J. Thomson focused on an extensive experimental research program with the aim of investigating positive rays in discharge tubes. The positive rays or what in German were known as *Kanalstrahlen* (canal rays) were first identified in 1886 by Eugen Goldstein and subsequently studied in great detail by Wilhelm Wien and other physicists. Wien may be best known for his work on blackbody radiation, which included the displacement law of 1894 and the Wien radiation law of 1896, and it was for this work that he was awarded the Nobel Prize in 1911. He was also a pioneer in electron theory and, what in the present context is more relevant, in the study of the mysterious positive rays that appeared under the same circumstances as the cathode rays but behaved very differently. As Wien was the first to demonstrate, the mass-to-charge ratio m/e of the positive particles emitted from the anode was at least thousand times greater than for the negative particles (electrons). Confusingly, while the m/e value was constant for cathode rays, for the positive rays it varied considerably with the kind of gas in the low-pressure tube. The nature of the positive particles was largely unknown except that physicists assumed that the rays mostly consisted of positively charged ions.

Thomson's original aim was to study positive rays in order to understand the nature of positive electricity, its distribution in the atom, and its difference from the electrons making up negative electricity. In 1907, Thomson concluded that the hydrogen ion H^+ was the true elementary positive charge rather than the hypothetical e^+ counterpart to the electron (Section 1.3.1). This line of work developed into a more chemically oriented study, which culminated in a monograph in 1913 titled *Rays of Positive Electricity*. Using special discharge tubes, where the positive rays were deflected by strong magnetic and electric fields, he found that the particles formed parabolas of constant m/e on a fluorescent screen or photographic plate. Thus, with oxygen he obtained three parabolas corresponding to O^+, O^{2+} and O_2^+, and with hydrogen H^+ and H_2^+ parabolas.

Much of Thomson's experimental work was done in collaboration with his research assistant Francis Aston, who after studies in organic chemistry had changed his research interests to radioactivity and experimental physics. While working under Thomson, he discovered in 1913 that neon exhibited two parabolas, a discovery which he interpreted as due to the neon isotopes Ne-20 and Ne-22. The concept of isotopy had just been introduced by Frederick Soddy, and Aston's finding was the first evidence for the isotopic nature of non-radioactive chemical elements. Thomson made the same observation as Aston, but only reluctantly accepted the isotope interpretation. For a while, he suggested that the mass-22 parabola might be due to the exotic neon compound NeH_2^+. Aston later improved Thomson's positive-ray apparatus, which in 1919 resulted in the first version of the mass spectroscope or spectrometer, one of the most important and versatile instruments in twentieth-century physical sciences. In 1922, Aston's work on mass spectroscopy was awarded the Nobel Prize, not in physics but in chemistry. His work was basically about the nature of chemical elements, and elements belonged traditionally to the domain of chemistry.

The early experiments with hydrogen revealed not only H^+ and H_2^+ but also a component with *m*/*e* three times that of the H^+ ion. What could it be? Thomson referred to the entity as X_3 or X_3^+ and argued that it was likely to be H_3^+, a hypothesis he first stated in 1911 after having considered and rejected the possibility of a new element with atomic weight 3. Such an element, supposed to be a halogen below fluorine, had been contemplated by Mendeleev in 1902, but Mendeleev's super-fluorine was strongly electronegative and highly reactive, properties not shared by Thomson's X_3. The idea of H_3, a modified form of hydrogen, had first been aired by the Czech chemist Bohuslav Brauner, who in 1895 suggested that it might explain some of the spectral lines in the recently discovered helium gas (Section 2.1.2). However, Brauner's hypothesis was short-lived and Thomson may not have been aware of it.

According to Thomson, not only had he discovered a new complex ion, he had also discovered the neutral molecule H_3, an allotropic form which he characterized as chemically inert and more stable than any other known allotropic form of an element. The remarkably stable gas even resisted decomposition by electrical discharges. If the molecule were real it should exhibit a characteristic optical spectrum, but despite many attempts Thomson was unable to obtain solid spectroscopic evidence for the enigmatic H_3 molecule. Moreover, he was fully aware that its stability seemed to conflict with generally accepted ideas of valence and the chemical bond. After all, hydrogen was universally admitted to be univalent, so how could three hydrogen atoms form a molecule? While ozone O_3 could be reconciled with standard views of valence, H_3 could not.

Despite these and other problems, Thomson was convinced that H_3 was a real and stable molecule, a conviction he kept to throughout his life. As late as 1934, the 78-year old physicist repeated his belief that what he had first obtained in 1911 by bombarding potassium hydroxide KOH with cathode rays was a permanent gas. "There is not the slightest doubt that the H_3 obtained in this way is stable and can exist uncharged," he asserted (Thomson, 1934). The original gas, he now suggested, might in part have consisted of HD molecules, that is, a combination of ordinary and heavy hydrogen or deuterium atoms (D). Whether in the H_3 or the HD form, at that time no one else shared his belief.

Thomson's claim of having detected H_3^+ and H_3 was well known, but at first received with some skepticism especially as it concerned the neutral molecule. On the other hand, in experiments with positive hydrogen rays of 1916 the Canadian-American physicist Arthur Dempster, a pioneer of mass spectrometry working at the University of Chicago, confirmed the presence of H_3^+, as did a series of later experiments. The neutral molecule was more problematic. Among the few early supporters was Aston, who used his new mass spectrograph to determine the masses of atomic and molecular hydrogen. Aston was convinced of the H_3 hypothesis and in his book *Isotopes* in 1923 he reported its mass to be 3.025 atomic mass units, compared to 1.008 for atomic hydrogen. He concluded that, "The mass deduced proves in a conclusive manner that the particle causing it [the mass-3 parabola] is a molecule of three hydrogen atoms" (Aston, 1923, p. 70).

Experiments apart, in papers published in 1913 and 1915, Johannes Stark considered the molecular structure of Thomson's H_3, which he believed was real and could be explained in terms of his own unorthodox view of valence. According to Stark, a triatomic ring of hydrogen atoms could be formed by three atoms and electrons held in an equilibrium position by electric lines of force. Stark's ring-like pictorial model shared some of the features of Thomson's model, but it failed to attract much chemical interest. More importantly, in 1919 Niels Bohr examined in a little known communication from the Nobel Institute in Stockholm the H_3 molecule from a theoretical point of view.

Bohr had first considered the constitution of H_3 four years earlier, but at the time without publishing his considerations. As mentioned in Section 1.2.4, Bohr was, from an early date, interested in Thomson's experiments, and in 1913 he had suggested that the supposed H_3 molecule might possibly be a superheavy isotope of hydrogen, H-3 or what was later called tritium. Bohr's 1919 model consisted of three electrons rotating in a common circular orbit, with one nucleus in the middle and the other two symmetrically displaced from it. Having concluded from energy considerations that the system was mechanically stable, he wrote: "This model may therefore possibly correspond to the molecule of a new modification of hydrogen, for the appearance of which under suitable conditions interesting evidence has been brought forward by Sir J. J. Thomson in his well-known experiments on positive rays" (Bohr, 1981, p. 476). As regards the formation of H_3 in an ionized hydrogen gas, he suggested the process $H^+ + H_2^- \rightarrow H_3$.

While Bohr's theoretical investigation offered some support for the H_3 hypothesis, he found that the corresponding positive ion was mechanically unstable, dissociating as $H_3^+ \rightarrow H_2 + H^+$. On the other hand, his calculations showed the negative ion H_3^- to be stable. Bohr's model calculations made sense within the framework of the old quantum theory, but they exerted very little influence on the further discussion of triatomic hydrogen and molecular constitution in general. He did not follow up the paper, which was his last one dealing with the structure of molecules (Section 1.2.5). Nor did other quantum physicists feel any need to consider the structure of triatomic hydrogen. They investigated H_2 and H_2^+ in great detail, but ignored H_3 and H_3^+. After all, since practically nothing was known about the properties of these more or less hypothetical molecules, calculations could not be meaningfully compared with experimental data.

By the late 1920s, the existence of the H_3^+ ion was no longer seriously contested, a result of several improved experiments made by American and German researchers who convincingly confirmed and extended the earlier investigations by Thomson, Aston, and Dempster. This work led to a consensus view of the main reaction responsible for the formation of H_3^+, namely, the collision of H_2^+ ions with ordinary hydrogen molecules according to $H_2^+ + H_2 \rightarrow H_3^+ + H$. Mass-spectroscopic measurements in the early 1930s left no doubt about the existence of H_3^+, which unmistakably turned up in the mass spectrum. After the discovery of deuterium, physicists also found hydrogen components of higher mass than three, such as H_2D^+, D_2H^+, and D_3^+.

Although H_3^+ was no longer controversial, almost nothing was known about its properties or molecular constitution. Several scientists looked for the optical spectrum of the ion, but with no success, and neither did complicated calculations based on quantum mechanics yield unambiguous information about its structure. Some calculations indicated a triangular structure, somewhat analogous to ozone, but it was uncertain whether or not the system had a stable configuration. Thus, although the triatomic hydrogen ion was generally accepted by the early 1930s, it remained an open question what it was, more precisely. This was disturbing, but not nearly as disturbing as the status of the gas supposedly made up of neutral H_3 molecules.

2.3.2. ACTIVE HYDROGEN

The question of the neutral H_3 molecule turned out to be much more complex and controversial than Thomson had anticipated. While some chemists and physicists were in favor of the hypothesis, others doubted it and still others rejected it. Those in favor often referred to "active hydrogen," a descriptive term that did not necessarily imply commitment to the H_3 hypothesis but only to some unusual form of polyatomic hydrogen. The possible existence of active hydrogen was a hot research topic in the 1920s, when it was investigated by dozens of scientists, mostly chemists but also some physicists.

The idea of a reactive form of molecular hydrogen was introduced by two scientists at Harvard University, William Duane and Gerald Wendt, who in 1917 reported experiments with a pure hydrogen gas intensely exposed to alpha rays. The modified form of hydrogen reacted with sulfur to produce hydrogen sulfide H_2S and with phosphorous to produce phosphine PH_3. It also attacked mercury and had a higher boiling point than ordinary hydrogen. Duane and Wendt further noticed a small volume contraction, less than 1%, which they speculated might be due to the formation of Thomson's triatomic hydrogen by a net process such as $3H_2 \rightarrow 2H_3$. However, the conclusion of the two Americans was cautious: "The chemical activity may be due to the formation of H_3 by the alpha rays. Certain characteristics of the active gas seem to point in this direction. We have not obtained, however, conclusive proof of the existence of H_3" (Duane and Wendt, 1917). In a later paper, Wendt and his student Robert Landauer found that H_3 was unstable with a life-time of the order of a minute. Contrary to the opinion of

Thomson, the new experiments suggested that the active gas was a hydrogen version of ozone:

> "The properties of the new gas are precisely those to be expected of an ozone form. It is much more reactive than ordinary hydrogen, yet not as active as nascent [atomic] hydrogen. Its boiling point is much higher than that of ordinary hydrogen. And it is unstable, for it cannot be detected if more than about a minute is allowed to elapse between the time of its formation and its reaction with a reducible substance. All the evidence obtained, then, points to the formation of triatomic hydrogen, perhaps properly called 'hyzone,' when even hydrogen is ionized" (Wendt and Landauer, 1920).

The name "hyzone" never caught on. It was occasionally used in the interwar period but soon fell into oblivion and is today nothing but a historical curiosity. Wendt and Landauer also confirmed the contraction effect supposedly due to $3H_2 \rightarrow 2H_3$, for which they reported a reduction in volume as high as 27%.

During the 1920s, much work was done on the active hydrogen, which in many but not all cases was thought to be identical to triatomic hydrogen. The cumulative result of many experiments was an increased confidence in H_3, yet without the growing evidence in favor of the molecule amounted to a recognized discovery. G. N. Lewis apparently accepted triatomic or active hydrogen, which he referred to in his influential book *Valence and the Structure of Atoms and Molecules* published in 1923. According to Lewis, H_3 did not have a ring structure but might be a loose combination between an H atom and a H_2 molecule. The H_3 hypothesis was taken seriously enough to enter a few textbooks, such as the British chemist Joseph Mellor's *Modern Inorganic Chemistry*, which in its edition of 1925 included a brief description of allotropic or triatomic hydrogen. The first PhD thesis on the subject, supervised by Wendt, was completed by Landauer in 1922, to be followed by several other theses related to the subject.

As mentioned, in 1913 Bohr had suggested that Thomson's triatomic hydrogen might possibly by an isotope of hydrogen with mass 3, a possibility which for a brief while played a role in the discussion concerning the reality of the molecule. In a paper from

1920, Wendt and Landauer referred to the possibility if only to reject it as unlikely. After all, it was hard to imagine how H-3 could be formed if the H_3 gas was the result of experiments with ordinary hydrogen. Nonetheless, the possible existence of two forms of H_3 — one a triatomic molecule and the other a mass-3 isotope of hydrogen — led occasionally to confusion. Part of the confusion was due to the American physical chemist William Harkins, who designated the hypothetical H-3 isotope H_3 and for the equally hypothetical H-2 isotope chose the symbol H_2. Thus, in a report of 1921 on inorganic chemistry the British chemist Edward Baly, at Liverpool University, stated: "Now there seems no doubt that the helium isotope discovered by Rutherford [He-3] is a different entity from H_3, which ... was first discovered by Thomson, now confirmed by Aston, and has recently been directly prepared by activation of hydrogen" (Baly, 1921). The report continued: "Aston has definitely shown that H_3 carries one charge, and this fact, considered along with its formation from hydrogen, shows that it is an isotope of hydrogen. There thus exist two elements of mass 3, one an isotope of hydrogen and the other an isotope of helium."

The experimental evidence cited in favor of H_3 was diverse. Thus, Duane and Wendt reported evidence that active hydrogen had a boiling point as much as 70 degrees above the one of ordinary hydrogen, an observation which in qualitative terms was confirmed by a few other chemists. The evidence also included the observation mentioned earlier of a volume contraction effect in a gas of H_2 when acted upon by alpha rays. However, the size and cause of the effect remained controversial. Whereas some experiments confirmed it, in other experiments it conspicuously failed to turn up.

All researchers active in the field agreed that the activity of the new hydrogen form quickly reduced, but the lifetime varied from one experiment to another. While some chemists found a lifetime at about two minutes, others concluded that it was less than a second, and there were even suggestions of a lifetime as short as 3×10^{-8} s, the same order as the life-time of ordinary excited molecules. An important chemical reaction specific to active hydrogen was its willingness to combine with nitrogen and form ammonia even at room temperature and without the help of catalysts. This reaction was sometimes used as a test for the active polyatomic hydrogen. The Canadian chemist Audrey Grubb suggested in 1926 that a possible reaction mechanism was two successive collisions of a N_2 molecule with H_3 groups, that is, $(N_2 + H_3) + H_3 \rightarrow 2NH_3$.

In addition to the reaction with nitrogen, triatomic hydrogen reacted with mercury, as Thomson had concluded in his early experiments. Duane and Wendt further found that the formed mercury hydride, possibly HgH_2, decomposed on heating. The most consistent and sensitive of the chemical tests for active hydrogen relied on the formation of hydrogen sulfide H_2S on a deposit of sulfur powder, with the produced H_2S being identified by its reaction with lead acetate on strips of paper. A gray or black precipitate of PbS on the paper indicates the presence of even very small amounts of hydrogen sulfide: $Pb(CH_3COO)_2 + H_2S \rightarrow PbS + 2CH_3COOH$.

The situation around 1925 was, then, that there was considerable evidence for the existence of H_3 in an unstable state, but also that the evidence was not very firm and lacked convincing confirmation. Spectroscopic identification remained a problem. Although a few advocates of active hydrogen argued to have found indirect spectroscopic evidence for the gas, it was too ambiguous to impress those who doubted the hypothesis. Within a few years, the balance shifted away from the advocates and to the side of the doubters.

2.3.3. COUNTEREVIDENCE AND DISBELIEF

While the existence of the H_3^+ ion enjoyed general acceptance by the mid-1920s, a result first and foremost caused by mass-spectroscopic measurements, the epistemic status of the neutral molecule was quite different. An increasing number of chemists and physicists began questioning the existence of H_3 in the form of active hydrogen. Together with the still missing spectroscopic evidence, the accumulation of experimental arguments against the elusive molecule had the effect that neutral triatomic hydrogen gradually lost its credibility. On the epistemic scale, it was downgraded from being a "likely" molecule to one "most unlikely."

The arguments against the H_3 hypothesis were primarily experimental and based on critical reexaminations of earlier studies in support of the hypothesis. Thus, in the late 1920s, several chemists pointed out that the evolution of hydrogen disulfide H_2S apparently caused by the contact of active hydrogen molecules with sulfur powder could be explained without assuming a new and highly reactive form of polyatomic hydrogen. Repeating the H_2S standard test for active hydrogen without using sulfur powder at all, or any other form of sulfur, a Russian chemist reported a positive result! He and other chemists argued that the H_2S observed in earlier experiments was due to small amounts of impurities in the H_2 gas used in the experiments.

Other chemists argued that several of the properties of active hydrogen could be explained in terms of monatomic hydrogen (H), the existence of which was well established at the time. According to Irving Langmuir, this form of hydrogen could be obtained by heating platinum or palladium to a high temperature in a gas of low pressure. Langmuir reported that the atomic hydrogen so prepared combined with oxygen at room temperature, and with phosphorous to form phosphine. Although monatomic hydrogen was even more reactive than active hydrogen, some of the effects ascribed to H_3 might be accounted for as effects caused by H atoms. For example, the small contraction of the H_2 gas, a dilatometric effect frequently cited in favor of the triatomic hydrogen molecule, could be explained as due to the adsorption of monatomic hydrogen on the glass walls of the apparatus.

Many of the critics of the H_3 hypothesis, among them Harold Urey and Fritz Paneth, phrased their objections cautiously. G. R. Schultze, a chemist at the University of Minnesota, was more direct in his conclusion, which he expressed in plain words:

> "Articles dealing with the existence of triatomic hydrogen appear over and over again. ... The more one becomes acquainted with these problems the more one realizes how little has actually been achieved toward establishing of concrete proof of the existence of triatomic hydrogen. ... I do not believe that a mere proof of the existence of some sort of activation is sufficient basis for the assumption of the existence of H_3 when one has a possible explanation by means of the well established atomic modification of active hydrogen. ... [It has been] shown conclusively that triatomic hydrogen does not exist ... [and] that the question of the non-existence of triatomic hydrogen has been definitely settled" (Schultze, 1931).

As a further reason to dismiss the H_3 hypothesis, Schultze referred to calculations of the chemical bond. As he pointed out, whereas Bohr's old theory allowed H_3, the new and much more reliable quantum chemistry based on the Heitler–London theory of the covalent bond did not. "Modern physics has furnished strong theoretical arguments against such a molecule," he stated. The molecular systems H_3 and H_3^+ attracted theoretical interest at an early data and were studied by several of the pioneers of quantum chemistry, such as Fritz London, Charles Coulson, Linus Pauling, and Henry Eyring.

Most of the theoretical work during the 1930s was concerned with the H_3^+ ion, which the new generation of quantum chemists considered important because it is the simplest possible triatomic molecular system, consisting of just three protons and two electrons. Coulson, who was soon to become a leader of quantum and computational chemistry, concluded in 1935 that while H_3^+ might exist in its ground state in the form of an equilateral triangle, the excited states would be unstable. As for the H_3 molecule, his calculations suggested that it was linear and highly unstable, immediately dissociating as $H_3 \rightarrow 3H$, a conclusion which also appeared in Coulson's important monograph *Valence* from 1951.

The most elaborate calculations of the early period were made by a research group led by Joseph Hirschfelder, a physical chemist at the University of Wisconsin. Hirschfelder concluded after lengthy computations that the H_3^+ ion had a stable configuration corresponding to a separation between the nuclei of about 1.8 Å with the nuclei lying intermediate between a right and an equilateral triangle. As a way of producing the triatomic ion, he predicted the reaction $H_2 + H_2^+ \rightarrow H_3^+ + H$, which according to his calculations was exothermic. Hirschfelder calculated the vibration frequencies of the equilateral form of H_3^+, finding that "the z and y modes of vibration should be infra-red active and therefore susceptible to direct experimental observation" (Hirschfelder, 1938). The remark was potentially important, but it was not followed up and did not lead to an experimental search for the spectral lines.

The complicated calculations of the quantum chemists did not unequivocally rule out the existence of H_3, but they offered little hope that the molecule could exist in a detectable form and hence be of chemical interest. It is also worth mentioning that from

Thomson's discovery claim in 1911 up to the early 1960s there were no suggestions that H_3 or H_3^+ might be of any interest to the astronomers, such that it appeared later on. During the three decades between 1940 and 1970, research in triatomic hydrogen came to a halt, with the average number of papers per year on the subject being less than one.

2.3.4. DISCOVERIES OF H_3 AND H_3^+

In 1961, at a time when molecular astrophysics was still in its infancy, three physicists at the Georgia Institute of Technology suggested in a brief note to *Astrophysical Journal* that H_3^+ might be abundant in interstellar space. They argued that the ion would be formed by $H_2 + H_2^+ \rightarrow H_3^+ + H$, the same reaction that had been considered by Hirschfelder many years earlier. The paper concluded: "It may be expected that H_2^+ will be converted to H_3^+ upon encounter with a hydrogen molecule, and the population of H_2^+ will be very strongly influenced by the density of neutral molecular hydrogen. It now appears desirable to consider the possibilities for detecting H_3^+ because this molecular ion may be present under some circumstances to the virtual exclusion of H_2^+" (Martin *et al.*, 1961). In retrospect, this was an important contribution to a new research field, but at the time the suggestion aroused very little attention. Until 1972, the paper of the three American physicists received only four citations, which means that its impact was close to zero.

The breakthrough came about 1980, when two researchers at the new Herzberg Institute of Astrophysics in Ottawa detected spectral lines from triatomic hydrogen and its associated positive ion.

Gerhard Herzberg, a German–Canadian specialist in molecular spectroscopy, was awarded the 1971 Nobel Prize in chemistry for "his contributions to the knowledge of electronic structure and geometry of molecules, particularly free radicals." He might just as well have received the prize in physics or, had there been such a prize, in astronomy. Originally trained in engineering physics at the Darmstadt Technical University, from where he graduated with a doctoral dissertation in 1928, Herzberg specialized in spectroscopy in which area he soon became recognized as an authority. In 1936, he published a classic work, which the following year was translated as *Atomic Spectra and Atomic Structure*. At the time of publication, he had left the Third Reich for Canada (his wife was Jewish) and begun focusing on astronomical applications of molecular spectra. Herzberg made his first important discovery in this area in 1941, when he and his graduate student Alex Douglas succeeded in reproducing unidentified spectral lines from the interstellar medium and showing that they belonged to the CH^+ ion. This was the first molecular ion found in interstellar space.

Since 1947, Herzberg worked in a research position at the National Research Council in Canada, which in 1975 established the Herzberg Institute of Astrophysics in recognition of his outstanding scientific contributions. Herzberg had, for long, had an interest in triatomic hydrogen, and had in vain looked for an absorption spectrum in the infrared region. He now decided to make a new attempt, this time to find the infrared emission spectrum of H_3^+ produced in a cathode discharge tube. He did not immediately succeed, but in early 1979 he realized that while searching for H_3^+ he had unexpectedly found discrete lines from H_3 in an excited state with one of the electrons being far from the three protons. As he recalled a few years later:

> "I decided to look for the emission spectrum of this ion [H_3^+]. At the same time in our laboratory T. Oka was looking for the absorption spectrum. After four years of hard work Oka finally observed the absorption spectrum of H_3^+, but my own search for the emission spectrum was unsuccessful. However, in this search I found a new spectrum, which turned out to be the spectrum of the neutral H_3. This was a case of sheer serendipity. I was not looking for H_3. It took me one or two months to recognize that the spectrum that we had observed was due to H_3 in Rydberg states, but once I realized this, everything fell into place" (Herzberg, 1985).

Seventy-five-year old Herzberg not only discovered lines from H_3 but also from its deuterium analog D_3. In his brief communication of March 1979 to *Journal of Chemical Physics* he reported "the observation of the Rydberg spectrum of an unstable polyatomic molecule, namely discrete spectra of triatomic hydrogen (H_3 and D_3)" (Herzberg, 1979). Herzberg found that the geometry of the molecules was an equilateral triangle, as expected from theoretical considerations. As mentioned, he realized that the observed triatomic molecules were in Rydberg states, the kind of highly excited states which in the case of atoms had first been suggested by Bohr back in 1913 (Section 1.2.3). At about the same time he also detected lines due to NH_4, another Rydberg molecule. Diatomic Rydberg molecules had been known for some time, but H_3 and NH_4 were the first polyatomic molecules of this kind.

Herzberg's discoveries in 1979 were quickly recognized as a breakthrough in molecular science. In 1994, on the occasion of his 90th birthday, he was asked in an interview for the most satisfying of his many discoveries. He answered that H_3 definitely belonged to his favorites, because it was quite unexpected. Herzberg's discovery of H_3 was important, but even more important was the succeeding discovery of H_3^+ made by Takeshi Oka, a Japanese–American astronomer and astrochemist who for most of twenty years worked with Herzberg at his institute in Ottawa. In friendly competition, the two scientists searched for H_3^+, with Herzberg looking for the emission spectrum and Oka for an absorption spectrum. In 1980, Oka finally succeeded in measuring fifteen infrared absorption lines of the H_3^+ spectrum, which had eluded all previous workers since the time of Thomson some seventy years earlier. While Herzberg's discovery of H_3 was serendipitous, Oka's discovery of H_3^+ was not.

In his discovery paper of 1980, Oka pointed out that his work opened new paths for future research, one of them being the detection of H_3^+ in space. Indeed, he immediately dropped his experiments and started to search for the infrared spectrum of the triatomic hydrogen ion that would reveal its existence in interstellar space. For several years, his efforts bore no fruit. The first detection of H_3^+ outside the laboratory was made in 1987, when spectral lines from the outer atmosphere of Jupiter were identified as belonging to the ion, and three years later Oka and his collaborator Thomas Geballe finally obtained evidence for the presence of H_3^+ in interstellar space.

It quickly turned out that H_3^+ was surprisingly ubiquitous, not only found in interstellar space and planetary atmospheres but also in the central area of the Milky Way. Looking back on his discovery of H_3^+ and what followed from it, Oka summarized:

> "My quest of interstellar H_3^+ has gone through three steps: laboratory spectroscopy from 1975 to 1980, the search for interstellar H_3^+ from 1981 to 1996, and the study of the Galactic center from 1997 till now. As a funny story goes, each went through three stages. When I was attempting, people said it was impossible. When I found it, people were skeptical. When it was all established and dust is settled, people said it was obvious. The first two discoveries have gone through the full cycles. They are obvious now. My on-going Galactic center work is in the middle" (Oka, 2011).

The work of Oka and his collaborators marked the beginning of a large-scale interdisciplinary research program, which today is cultivated by hundreds of physicists, chemists, and astronomers. The simple triatomic hydrogen ion is more than just one out of many. It is one of the most abundant in the universe. In the twenty-first century, the study of H_3^+ has provided astrochemistry with a new dimension that could not possibly have been foreseen by pioneers such as Lockyer, Crookes, and Nicholson. Nor could it have been foreseen by Thomson, the father of the H_3^+ hypothesis.

2.4

Between Chemistry and Physics

2.4

Between Chemistry and Physics

The conventional distinction between chemistry and physics, or between chemists and physicists, is useful and perhaps inevitable, but it is much too coarse-grained to accord with the history of the two sciences. Scientists may be classified as physicists and yet be best known for their work in chemistry, or vice versa. The Nobel Prize system provides an illustration. It may seem natural that chemists are awarded prizes in chemistry and physicists in physics, but a large number of Nobel chemistry prizes have in fact been awarded to scientists who were either physicists or whose prize-winning work would be normally classified as physics. In this respect, the relationship is highly asymmetric, as no physics Nobel Prize has ever been awarded to a chemist or for a work belonging to chemistry as usually understood.

Radioactivity, isotopes, molecular structure, and the discovery of new chemical elements were and still are traditionally regarded as belonging to chemistry, yet many of the prizes in these areas have gone to physicists. An early and well-known example is the chemistry prize awarded to Rutherford in 1908. Rutherford considered his work on radioactivity to be strictly physical and held chemistry in low esteem. Having been informed of the prize, he wrote to Otto Hahn, "I must confess that it was very unexpected and I am very startled at my metamorphosis into a chemist" (Eve, 1939, p. 183). Three years later Marie Curie, also a physicist, was awarded the chemistry prize for her work on radium. Again, the 1951 Nobel Prize in chemistry was awarded to Glenn Seaborg and Edwin McMillan for their discoveries of the first transuranic elements (Section 2.5.2). While Seaborg was trained in chemistry, McMillan was an experimental nuclear physicist, such as he pointed out in his Nobel lecture. "In spite of what the Nobel Prize Committee may think, I am not a chemist," he said (McMillan, 1951).

The relationship between chemistry and physics has changed significantly over time, since the mid-nineteenth century with an increasing "physicalization" of chemistry. With physicalization followed a degree of mathematization. During the early phase, thermodynamics and statistical mechanics were particularly important, and during the later period quantum theory and atomic theory changed the chemical landscape. The impact of physical theories on chemistry, together with the general trend of professionalization and specialization in science, led to increased separation and sometimes to frictions over disciplinary boundaries, but it also led to progressive cross-fertilization and to the emergence of new interdisciplinary branches of research such as physical chemistry and chemical physics.

Moreover, there have always been scientists who with ease crossed the chemistry–physics boundary and contributed importantly to both fields. Michael Faraday may be best known as a brilliant physicist, but he also discovered benzene and the electrochemical laws named after him. To mention but one more scientist from the early period, the Austrian chemist Josef Loschmidt did important work in organic chemistry and was the first to estimate the size of gas molecules, but his work on kinetic gas theory and thermodynamics was no less important. Could an individual not span over both sciences, he or she could enter a collaborative partnership. Noteworthy examples from the nineteenth century include Bunsen and Kirchhoff (spectroscopy, 1860) and Ramsay and Rayleigh (argon, 1894).

2.4.1. FROM KANT TO NERNST

The relationship between chemistry and physics has always been a delicate one, for a long time with physics being considered a more noble and scientific area than the supposedly primitive and empirical chemistry. Under the spell of Newton's successful mechanical physics, Immanuel Kant argued in 1786 that chemistry could never be a genuine science because its subject matter was intractable to the mathematical method and systematic deduction from higher principles. "All proper natural science requires a pure part lying at the basis of the empirical part," the famous philosopher declared (Kant, 2004, p. 6). He could find no such pure part in chemistry, an art where the results did not follow with necessity from the laws of nature, such as did the results from the admired mechanical physics. Here is Kant's verdict on the non-scientific nature of chemistry:

> Chemistry can be nothing more than a systematic art or experimental doctrine, but never a proper science, because its principles are merely empirical, and allow of no a priori presentation in intuition. Consequently, they do not in the least make the principles of chemical appearances conceivable with respect to their possibility, for they are not receptive to the application of mathematics.

Kant wrote at a crucial turning point in the history of chemistry, just at the time when Lavoisier had abolished phlogiston and completed his chemical revolution based on precise measurements and advanced instruments. But Lavoisier's new chemistry was no less experimental that the earlier one, and no more mathematical. It did not impress the critical-idealistic philosopher from Königsberg.

It is important to realize that when Kant, or Lavoisier for that matter, spoke of chemistry they did not refer to what we call chemistry today. The meanings of the concepts chemistry and physics, and the relationships between them, have changed significantly over time. In the early part of the nineteenth century, it was customary to classify topics such as electricity, magnetism, heat, and light as chemical rather than physical sciences. Half a century later, the topics had become parts of physics or rather of what was then called "chemical physics." According to an American textbook published in 1874, "Chemistry is ... usually divided into two portions. The first treats of the Chemical Agents, Heat, Light, and Electricity, and is commonly called Chemical Physics; the second, of the chemical properties and relations of the various kinds of matter" (Pynchon, 1874, p. 21). What at the time was known as chemical physics was something very different from the modern meaning of the term, which refers to an interdisciplinary branch of science originally based on quantum chemistry and molecular spectroscopy (Section 2.4.4).

The connection and later disconnection between physics and chemistry in the nineteenth century can be followed by the content and titles of scientific textbooks and journals. One of the most important journals was *Annalen der Physik* founded in 1799 and later appearing under other names. During the period 1819–1824, its name was *Annalen der Physik und der Physikalischen Chemie* after which it changed to *Annalen der Physik und Chemie*. As indicated by its title, the journal was meant to be a medium for both physicists and chemists, but with the increasing specialization it became more and more physics-oriented and less chemistry-oriented. At around 1870, it was in reality a physics journal, but only in 1900 did the editors take the consequence, returning to the old title *Annalen der Physik*. By that time, frontier research in either physics or chemistry required its own journals. In France, we witness a similar development. The group around Lavoisier founded *Annales de Chimie* in 1789, which changed to *Annales de Chimie et de Physique* in 1815, and in 1914 was split up in two separate journals, *Annales de Chimie* and *Annales de Physique*.

In the early part of the nineteenth century, there were several potential meeting points of physics and chemistry. One of them was galvanism, and more specifically the explanation of the electric voltage of a battery. While Alessandro Volta and later Georg Ohm defended the physical contact theory, according to other scientists the action of Volta's battery was due to chemical reactions. Was the battery to be explained in terms of physics or of chemistry? The question led to an extended controversy that involved several of the period's prominent physicists and chemists.

Another meeting point was the concept of "affinity," the poorly defined force or power that somehow governed the willingness of two elements to unite or the ease at which a binary compound decomposed. Was affinity an electrical force? Or perhaps it was gravitational in nature? Following the emergence of the law of energy conservation, Julius Thomsen in Denmark and Marcellin Berthelot in France proposed an ambitious thermal theory of affinity, namely, that it was given by the amount of heat evolved in a chemical process. However, the otherwise progressive thermochemical research program was faced with a number of anomalies, the most serious one being the existence of spontaneous endothermic processes such as $N_2 + O_2 \rightarrow 2\ NO$. By the mid-1870s, it became increasingly clear that to explain chemical reactions one needed not only the first law of thermodynamics but also the second law. Energy could not stand alone, it had to be supplemented with entropy (Section 1.1.3). The result was the emergence of chemical thermodynamics, a highly abstract yet empirically powerful theory principally developed by Josiah Willard Gibbs and Hermann Helmholtz in the decade between 1873 and 1883.

None of the two founders of chemical thermodynamics were trained chemists. Their difficult works were more addressed to physicists than to chemists, most of whom considered the new theory to be too abstract to be of any relevance to practical chemistry or to enter chemistry textbooks. Although Helmholtz was greatly interested in the theoretical foundation of chemistry, he had but a low opinion of the kind of laboratory work that occupied the scientific life of the majority of chemists. As he somewhat arrogantly expressed it in a letter of 1891: "Thermodynamic laws in their abstract form can only be grasped by rigorously schooled mathematicians, and are accordingly scarcely accessible to the people who want to do experiments on solutions and their vapor tensions, freezing points, heats of solution, &c." (Kragh, 1993, p. 429). This was the voice of a theoretical physicist, not a chemist.

Once recognition had been accorded to the theories of Gibbs and Helmholtz, chemistry had a solid thermodynamic foundation, which was further developed by theoretical chemists and chemically oriented physicists. Among those who developed the Gibbs–Helmholtz thermodynamics, Max Planck was among the most active. Planck is of course famous for his work on blackbody radiation, which in 1900 led him to the quantum hypothesis, but during the 1890s his main field of study was chemical thermodynamics, an area which had in common with blackbody radiation that entropy features as a central concept. (It was considerations of the entropy of blackbody radiation that led Planck to the quantum hypothesis.) It is probably not well known that young Einstein was yet another physicist engaged in work on thermodynamics in general and entropy in particular. His earliest papers, inspired by Helmholtz, Gibbs, and Boltzmann, focused on the relations between thermodynamic macroscopic bodies and molecular-theoretical microscopic quantities. They belonged as much to physical chemistry as to pure physics.

Chemical thermodynamics was one of the pillars on which the new physical chemistry was based. Established as a discipline of chemistry in 1887, the year of the founding of *Zeitschrift für Physikalische Chemie*, the principal architects of physical chemistry were Wilhelm Ostwald, Svante Arrhenius, Jacobus van't Hoff, and Walther Nernst. There was among the early generation of physical chemists a discussion of the role and status of thermodynamics relative to the role of osmotic pressure, another of the pillars (the third of the pillars was ionic dissociation). Thus, Nernst got involved in a minor controversy with the Dutch chemist Johannes van Laar, who was a specialist in chemical thermodynamics and an advocate of the abstract entropy-based approach of Gibbs and Planck that Nernst disliked. Van Laar saw this approach as the royal road toward a "mathematical chemistry," as he called it. He argued his case in a book from 1901 with the remarkable title *Lehrbuch der Mathematischen Chemie* (Textbook on Mathematical Chemistry). Had Kant known about van Laar's book, he might have rejoiced.

Although chemical thermodynamics was a major step toward integrating physics and chemistry, it was at most a partial and in itself not a very convincing integration. By its very nature, thermodynamics is concerned only with state functions and bulk matter, whereas it has nothing to say about the microstructure of matter or the chemical elements and compounds that laboratory chemistry is first and foremost about.

2.4.2. WHAT IS A CHEMICAL ELEMENT?

As pointed out in Section 2.1.3, in the early years of the twentieth century Mendeleev resisted the developments in physics which indicated that atoms were composite bodies and that they might in some cases, as in radioactive elements, change from one kind of atom to another. As he saw it, what was at stake was the very foundation of chemistry, namely, the doctrine of immutable chemical elements defined by their atomic weights. Mendeleev was far from the only chemist who disliked the intrusion of physics into chemistry, whether in the form of mathematical methods, composite atoms, ions, electrons, or radioactive disintegration. Consider Arthur Smithells, a professor of chemistry at the University of Leeds, who at the 1907 meeting of the British Association expressed his worries about the problems that had come from recent developments in physics. He lamented "the invasion of chemistry by mathematics and, in particular, from the sudden appearance of the subject of radio-activity with its new methods, new instruments, and especially with its accompaniment of speculative philosophy" (Smithells, 1907). Smithells was tempted to see radiochemistry as "a chemistry of phantoms" and deplored the "exuberance of mathematical speculation of the most bewildering kind concerning the nature, or perhaps I should say the want of nature, of matter." Referring to the chemists of the older school, himself included, he wondered "whether chemistry was not beginning, as it were, to drift away from them."

Although a considerable part of the chemical community, and not least the large fraction consisting of organic chemists, was sympathetic to views like those expressed by Mendeleev and Smithells, far from all chemists feared the new sciences of radioactivity and atomic structure. In the early phase of radioactivity, chemists played as important a role as the physicists did. In the 1910s, the challenges that radioactivity and atomic theory caused to the traditional notion of a chemical element led to a new consensus view as to what constitutes an element. The change was conservative insofar as it preserved the status of the older elements, but otherwise it was radical because it transferred the definition of an element from the chemists' atom to the physicists' atomic nucleus.

In about 1911, Soddy suggested that substances with different atomic weights and radioactive properties might be conceived as chemically identical and hence belonging to the same element. Two years later, he proposed the word "isotope," which he related to Rutherford's new nuclear model of the atom (Section 1.2.4). The Dutch amateur physicist Antonius van den Broek had slightly earlier proposed disconnecting the ordinal number of the periodic table from the elements' atomic weight and instead identifying it with the nuclear charge. In effect, van den Broek introduced the concept of an atomic number, although he did not refer to the term, which may have been coined by Rutherford.

Importantly, on the basis of Bohr's theory, 26-year-old Henry Moseley realized that the atomic number could be determined by means of X-ray spectroscopy, an immensely useful method that he pioneered in investigations starting in 1913. Moseley established that the wavelength of any characteristic X-ray line decreased regularly with the ordinal number of the element in the periodic table. With $\bar{\nu}$ denoting the wave number $1/\lambda$ and R Rydberg's constant, he found that for the strong K_α lines,

$$\bar{\nu} = \frac{3}{4}R(Z-1)^2 \quad \text{or} \quad \sqrt{\bar{\nu}/R} = \sqrt{3/4}(Z-1)$$

a formula often known as Moseley's law. In accordance with Bohr, he interpreted the factor 3/4 as $3/4 = 1/1^2 - 1/2^2$ (Section 1.2.1). For other X-ray series designated K_β, L_α, and L_β Moseley obtained similar linear relationships. Thus, by measuring the X-ray wavelengths one could establish the atomic number and hence the position in the periodic table for all elements including the elusive rare earth metals. The X-ray spectroscopic method was created by a physicist and further developed by other physicists, in particular Maurice de Broglie in France and Manne Siegbahn in Sweden. Although most of the results flowing from the method were concerned with elements and their place in the periodic system, and hence belonged to chemistry, at first chemists took little notice of the new method.

What matters is that with the introduction of the twin concepts of isotopy and atomic number, the time was ripe for a reconsideration of the true nature of an element. But the transition from the old to the new definition of an element did not come easily, for other reasons because it was resisted by parts of the chemical community. Another reason was the First World War, which made scientific communication between the belligerent parties nearly impossible. It took until the early 1920s before the new definition received official recognition.

The Austrian physical chemist Fritz Paneth was a specialist in radiochemistry and instrumental in advocating and clarifying the new definition based on the atomic number. In a paper from 1916, he repeated what Soddy had stated earlier, namely that an element is a substance in which all the atoms have the same nuclear charge. Recognizing that the nuclear charge alias the atomic number might be a concept foreign to most of his colleagues in chemistry, he offered an alternative and more operational definition supposedly more agreeable to traditional chemists: "An element is a substance which cannot be decomposed by chemical means. Substances which satisfy this definition, count as one and the same element if they, once mixed with one another, cannot be separated by chemical means" (Paneth, 1916). Paneth thus stated that although a mixture of isotopes was physically inhomogeneous, it was homogeneous from a chemical point of view.

At first, Paneth's definition was met with considerable resistance from German-speaking chemists, many of whom found that it was foreign to the chemical mind. Nonetheless, in 1921, the German Atomic Weight Commission decided to base its new table of elements on the atomic number and Paneth's definition. The German commission was officially part of the International Committee on Atomic Weights established in 1899, but during the war international collaboration broke down and after the war the Germans were excluded from what now became the International Committee on Chemical Elements, a section of the new International Union of Pure and Applied Chemistry better known as IUPAC. Only in 1923 did the International Committee decide that the atomic number was the defining parameter of an element. The previous year, the Solvay Institute of Chemistry had held its first meeting in Brussels, without the participation of German and Austrian chemists. The question of what constituted an element was eagerly discussed in Brussels, with several chemists complaining that the atomic number made less sense from a chemical point of view than did the well-known atomic weight.

More than anything else, Moseley's powerful method of X-ray spectroscopy proved in practice the superiority of the new definition of an element. The first element ever to be identified directly from its atomic number as revealed by its X-ray spectrum was hafnium, element 72 in the periodic system. The discovery process evolved into a major controversy between scientists from Paris, who claimed to have discovered the element in rare earth metals, and scientists from Copenhagen, who found its characteristic X-ray lines in zirconium minerals in agreement with Bohr's theory of the periodic system (Section 1.2.5). The two Frenchmen, the chemist Georges Urbain and the physicist Alexandre Dauvillier, claimed to have detected very weak X-ray lines of what they called "celtium" (Ct) belonging to the rare earths. Their discovery claim in June 1922 was questioned by George Hevesy and Dirk Coster, working in Bohr's laboratory in Copenhagen, who in January 1923 announced the discovery of "hafnium" (Hf), a chemical homolog of zirconium. Whereas the optical spectrum of Hf was quickly obtained, it never turned up for Ct.

Although the evidence for Hf was much stronger than for Ct, the rival claims caused a long and unpleasant controversy that was only settled after several years of discussion. The International Committee refused throughout the 1920s to give official credit to Hf and only recognized it in 1930. During the controversy, support for Ct came almost exclusively from the quarters of traditional chemists, who considered the discovery of new elements a task for chemistry and not physics. The Czech chemist Bohuslav Brauner, an expert in rare earth chemistry and an ardent supporter of Ct, belonged to this group. He accused the Copenhagen team of lacking chemical skills and relying one-sidedly on physical methods. As he put it, "Only a chemist who is a specialist in this department [rare earths] is able to see that splendid physicists may show a great want of a truly 'chemical sense'" (Brauner, 1923). The tension between traditional chemistry and the new physics was an important element in the Hf–Ct controversy, which included a boundary conflict between professional interests. Many chemists felt it intolerable that outsiders, young physicists with hardly any chemical knowledge at all, arrogantly claimed to settle an important chemical problem. It was neither the first nor the last time that the tension between the two sister sciences came out in the open. A more recent example is the dispute between chemists and physicists regarding the discoveries of artificial elements near the end of the periodic system (Chapter 2.5).

2.4.3. STRONG ELECTROLYTES

What Smithells in 1907 had described as the physicists' "invasion of chemistry" accelerated over the following decades. The invasion took many forms and was sometimes phrased in an imperialist rhetoric by the conquering physicists. Thus, in 1920, Max Born reflected on the relations between the two sciences: "We realize that we have not yet penetrated far into the vast territory of chemistry," he wrote, "yet we have travelled far enough to see before us in the distance the passes that must be traversed before physics can impose her laws upon her neighbor science" (Born, 1920). Although Bohr did not use Born's militant rhetoric, he agreed in the reductionist goal of deriving all chemical properties from calculations of atomic physics.

Several of the new generation of quantum physicists shared the view of chemistry as an immature field that could only be turned into a proper science by adopting the advanced methods of theoretical physics. They suggested that it would only be a matter of years until chemistry was completely mathematized and thus become truly scientific in Kant's sense. For a while, they thought that Bohr, with his quantum theory of atoms and molecules, might become the new Newton who succeeded in basing chemistry deductively on the higher principles of physics. According to the physicist Oliver Lodge, "we are living in the dawn of a kind of atomic astronomy which looks as if it were going to do for Chemistry what Newton did for the Solar System" (Lodge, 1924, p. 203). But this is not what happened.

While many of the physics intruders fought with the weapons of atomic and quantum theory, in other cases well-tested classical physics sufficed, such as the case with the important theory of electrolytes developed by Peter Debye and his assistant Erich Hückel in Zürich in the early 1920s. This theory relied on electrodynamics, thermodynamics, and statistical physics, but neither on quantum theory nor models of atomic structure.

According to Svante Arrhenius' important ionic theory first formulated in 1887, electrolytes in a solution were incompletely dissociated into ions in an equilibrium process of the form $AB \leftrightarrows A^+ + B^-$. The degree of dissociation α depended on the total concentration c as given by Wilhelm Ostwald's dilution law stated in 1888, namely

$$\frac{c\alpha^2}{1-\alpha} = K,$$

where K is the dissociation constant characteristic for the electrolyte AB. Since the fraction α is the ratio of the electric conductance Λ to the conductance at infinite dilution Λ_0, Ostwald's law can be written in terms of conductivities. For highly diluted solutions ($\alpha = \Lambda/\Lambda_0 \approx 1$), the result comes out as a linear decrease with the concentration:

$$\Lambda(c) = \Lambda_0 - kc.$$

Although at first controversial, by 1900 the Arrhenius–Ostwald theory of electrolytic dissociation was generally accepted and considered a cornerstone of the new progressive and authoritative physical chemistry. Measurements of the conductivity of strong electrolytes such as $BaCl_2$ and HCl had long showed that the constant K varied greatly with the concentration, but for about two decades the severe anomaly was largely ignored. In 1909 and again in 1919, the Danish chemist Niels Bjerrum argued from spectroscopic evidence that strong electrolytes were completely dissociated and that interionic forces had to be taken into account to explain experimental data. However, his arguments were dismissed by Arrhenius and for several years they failed to win support. The anomaly of strong electrolytes was only fully recognized at about 1920, and with the recognition followed the need to establish a theoretical explanation.

In two papers from 1923, Debye and Hückel provided an impressively complete and detailed theory of strong electrolytes in dilute solutions. The site of the publications was *Physikalische Zeitschrift* (and not a chemistry journal) of which Debye was the senior editor and Hückel on the editorial staff. The basic idea of the two physicists was that the distribution of ions is not completely random, but that, on average, an ion is surrounded by an oppositely charged ionic cloud or atmosphere. Based on this picture, they arrived after a series of abstruse calculations at a number of definite and testable results in much better agreement with experiments than previous theories. First of all, Debye and Hückel derived a formula for the variation of the electric conductivity with concentration, which for highly dilute solutions gave the expression

$$\Lambda(c) = \Lambda_0 - a\sqrt{c},$$

where a is a constant. Contrary to the linear expression based on the Arrhenius–Ostwald theory, the square root expression agreed convincingly with measurements. Despite its intimidating complexity, the Debye–Hückel theory was eagerly taken up by mathematically competent chemists, and within a few years it was recognized as an important breakthrough in physical chemistry. Yet not all chemists were happy with the theory. Arrhenius preferred to ignore it, resisting until his death in 1927 the new picture of strong and completely dissociated electrolytes. While Arrhenius' ionic theory was revolutionary in the late nineteenth century, he ended as a conservative when the theory was completed by a later generation.

The Debye–Hückel theory was the beginning of a new research program in physical chemistry and not the final word on solutions of strong electrolytes. The most important of the early extensions of the theory was due to 23-year-old Lars Onsager, a Norwegian student who had not yet graduated from the Technical University in Trondheim, Norway. In a paper published in 1926, Onsager pointed out some weaknesses in the Debye–Hückel theory and demonstrated that it could be improved by taking into account the Brownian motion of the ions brought about by the collisions of the surrounding solvent molecules. The Debye–Hückel–Onsager theory, as it is often called, resulted in a more general and empirically superior expression for the conductance equation $\Lambda = \Lambda(c)$. Onsager's work was the first step in a glorious career in theoretical chemistry which forty-two years later would be crowned with a Nobel Prize.

Hückel never received any formal training in chemistry, and his knowledge of the field was restricted to scattered reading of his own. Nor had Debye ever worked in a chemistry laboratory. His doctoral degree was in electrical engineering, and he subsequently excelled in a series of important theoretical investigations covering electromagnetic waves, quantum theory of solids, X-ray diffraction, and much more. Debye's first encounter with chemistry was his work on dilute electrolytes, after which he increasingly oriented his research toward the physics–chemistry interface. He was awarded the 1936 Nobel Prize in chemistry for his "contributions to our knowledge of molecular structure through his investigations on dipole moments and on the diffraction of X-rays and electrons in gases."

As far as Hückel is concerned, after having mastered the new quantum mechanics he turned on Bohr's advice to problems of quantum chemistry. In important works from the early 1930s, he studied the double bond and provided a quantum-mechanical explanation of the properties of the benzene molecule C_6H_6 and aromatic compounds in general. Hückel showed that the properties of aromatic compounds were due to a group of six electrons forming a complete electron shell and not, as traditionally believed, to six carbon atoms forming a ring. What is currently known as the Hückel method or Hückel molecular orbital (HMO) theory proved to be of great importance to computational chemistry related to organic molecules.

2.4.4. TOWARDS QUANTUM CHEMISTRY

In 1929, Paul Dirac was busy with analyzing the amazing consequences of his new relativistic theory of the electron (Section 1.3.2), but he also found time to deal with other lines of work. In the introduction to a paper on the quantum mechanics of many-electron atoms, he offered his reductionist view on the relationship between physics and chemistry:

> "The general theory of quantum mechanics is now almost complete, the imperfections that still remain being in connection with the exact fitting in of the theory with relativity ideas. ... The underlying physical laws necessary for the mathematical theory of a large part of physics and the whole of chemistry are thus completely known, and the difficulty is only that the exact application of these laws leads to equations much too complicated to be soluble" (Dirac, 1929).

In other words, Dirac suggested that with appropriate approximation methods chemistry would turn out to be nothing but applied quantum physics. As we have seen, this kind of attitude was not new as it had been part of the physicists' culture for several decades. With the rise of quantum mechanics, it appeared in an enhanced and more confident form, initially with the result that the gulf between physics and chemistry widened.

Molecules and the chemical bond keeping atoms together received attention in the old quantum theory, but despite much work the overall result was discouraging. The hydrogen molecule H_2 was examined in detail, first by Bohr and subsequently by Born and other physicists, yet the calculations failed to reproduce experimental data (Section 1.2.5). The even simpler case of H_2^+ consisting of only three particles fared no better. Pauli's elaborate theory of the H_2^+ ion was a failure, which suggested that existing quantum theory was unable to explain the structure of molecules. Most chemists preferred pictorial models of the kind proposed by G. N. Lewis, where a pair of fixed electrons provided the chemical bond between two atoms. Although Lewis' models flatly disagreed with established quantum theory, they were useful and made chemical sense.

The first application of the new quantum mechanics to what in principle was a chemical problem, the calculation of the bond strength in the H_2^+ ion on the basis of the Schrödinger equation, was made in 1926 by Øyvind Burrau at Bohr's institute. More important than Burrau's exercise in wave mechanics, the following year the two German physicists Walther Heitler and Fritz London attacked the H_2 molecule in a seminal work published in *Zeitschrift für Physik*. The paper is traditionally seen as the beginning of modern quantum chemistry. The basic approach of Heitler and London was to consider separately one of the electrons in each of the combining atoms and then, by clever use of approximation methods, to construct a wave function representing the paired-electron bond between them. Relying on Heisenberg's recent discovery of the so-called resonance effect, they showed that when the two electrons have opposite spins they tend to aggregate in the region between the two protons, thereby reducing the total energy of the system. The equations of Heitler and London gave promising agreement with experiments, namely a binding energy of the hydrogen molecule of 54 kcal/mole or approximately two-thirds of the one found experimentally. For the proton–proton distance, they obtained 0.8 Å. Apart from explaining the formation of the molecule from two hydrogen atoms, they also explained why two helium atoms cannot form a He_2 molecule.

The pioneering work of Heitler and London suggested a mathematization of chemistry more real and profound than what previously had been dreamt of. The American solid-state physicist John Van Vleck, who also contributed importantly to quantum chemistry, was inspired to ask: "Is it too optimistic to hazard the opinion that perhaps this is the beginning of a science of 'mathematical chemistry' in which chemical heats of reaction are calculated by quantum mechanics just as are spectroscopic frequencies of the physicist?" (Van Vleck, 1928). Van Vleck's rhetorical question appeared in a review article on quantum mechanics addressed to a chemical audience. From a slightly different perspective, the Heitler–London theory can be seen as yet another example of the "invasion" of a chemical territory by theoretical physicists with basically no background or interest in chemistry. London's lack of respect for the culture of chemistry may be glimpsed from a letter he sent to Heitler in 1935: "The word 'valence' means for the chemist ... a substitute for these forces whose aim is to free him from the necessity to proceed, in complicated cases, by calculations deep into the model. ... The chemist is made out of hard wood and he needs to have rules even if they are incomprehensible" (Gavroglu and Simões, 2012, p. 100).

The main result of the Heitler–London paper was its deductive argument that the covalent bond can be explained purely in terms of spin quantum mechanics and was therefore outside the reach of classical chemistry. As Heitler and London emphasized, spin and the associated exclusion principle were crucial in understanding the chemical bond and they belonged clearly to quantum physics, not to chemistry. The approach of the two physicists formed the backbone of what eventually became known as the valence bond (VB) method, which in the version developed by Linus Pauling and others dominated quantum chemistry during the 1930s.

Young Pauling was instrumental in making the chemical community aware of the relevance of the quantum physicists' treatment of simple molecular systems. In 1928, he wrote an extensive essay review for *Chemical Reviews* in which he explained to chemists the essence of Burrau's H_2^+ calculations and the

Heitler–London theory without using the advanced mathematics of quantum mechanics. Given Pauling's background in chemistry and later reputation as a "chemical translator," it is noteworthy that at the time he subscribed to the reductionist view expressed by Dirac and some other physicists. In an unpublished lecture from 1928 to a section of the American Chemical Society, he stressed that chemistry must necessarily follow the results of theoretical physics: "We can say, and partially vindicate the assertion, that the whole of chemistry depends essentially upon two fundamental phenomena: these are (1) the one described as the Pauli Exclusion Principle; and (2) the Heisenberg–Dirac Resonance Phenomenon" (Park, 1999).

In the same year that Heitler and London published their theory, Friedrich Hund laid the foundation of an alternative and very different way of conceiving the chemical bond. Inspired by Burrau's calculation of the H_2^+ ion, Hund assumed that an individual electron moved in the field from all the nuclei and the other electrons in the molecule. His approach was essentially to consider a diatomic molecule such as H_2 in its two most extreme and unrealistic states, one with the atoms completely separate and the other with the atoms completely united into a single system, and then find the intermediate distance at which the terms from atomic spectroscopy coincided with those of molecular spectroscopy. Compared to the Heitler–London method, Hund's method overemphasized the ionic character of the molecule. For example, in the H_2 case it assumed that the probability of having two electrons around the same proton was the same as having one electron around each of the protons.

Hund's work was paralleled by works done by Robert Mulliken, an American specialist in molecular spectroscopy who in 1926 had discovered the zero-point energy in molecules (see Section 4.4.2) and forty years later would be awarded the Nobel Prize for his important contributions to quantum chemistry. Contrary to Hund, who knew very little chemistry, Mulliken had a solid chemical training and a PhD in physical chemistry. In 1928, the theories of Hund and Mulliken resulted in the first version of what came to be the molecular orbital (MO) method, initially as a theory of molecular spectroscopy and not of valence. Hund and Mulliken turned their interest toward the chemical bond a little later, and only then was the molecular orbital theory or method seen as an alternative to the Heitler–London–Pauling valence bond method. According to Mulliken, the theory behind the latter method was unsatisfactory and its agreement with chemical data largely accidental. A basic feature of the Hund–Mulliken theory was the heuristic use of the so-called *Aufbauprinzip* (construction principle), which goes back to Bohr's theory of the periodic system (Section 1.2.5) and according to which the building-up of atomic orbits takes place by a series of successive captures of electrons in their lowest energy states. Hund and Mulliken made use of a somewhat similar molecular *Aufbauprinzip* for diatomic molecules. Although there were no classical electron orbits in the new theory, by means of the *Aufbauprinzip* the orbits could be transferred to molecules in the form of molecular orbitals.

As the year 1932 has been called the the *annus mirabilis* of nuclear physics, so the year 1931 has been called the *annus mirabilis* of quantum chemistry: it was in this year that the two methods, valence bond and molecular orbitals, reached maturity and that quantum chemistry entered as a social and scientific reality, an autonomous subdiscipline.

However, the new interdisciplinary theory percolated slowly into chemistry from its origin in quantum mechanics and molecular physics, and more slowly in the case of the molecular orbital than the valence bond method. It was mainly due to the efforts of Pauling that the chemical community discovered and came to appreciate quantum chemistry as an indispensable part of theoretical chemistry. Pauling sold the message of the quantum valence bond to the chemists in an influential series of papers published in 1931–1932 in *Journal of the American Chemical Society* with the common title "The Nature of the Chemical Bond," and in 1939 he published a classic and no less influential monograph with the same title. The book was written in a language that most chemists could understand, with only a minimum of mathematical formalism. In the preface, Pauling stressed that although quantum mechanics constituted the foundation of quantum chemistry, the new science was nonetheless chemical in nature: "A small part only of the body of contributions of quantum mechanics to chemistry has been purely quantum-mechanical in character. ... The advances which have been made have been in the main the result of essentially chemical arguments" (Pauling, 1939, p. vii). This was a view completely different from the one he had pronounced a decade earlier.

Pauling's *The Nature of the Chemical Bond* was dedicated to Lewis, whose early works on the structure of atoms and molecules Pauling much appreciated. He conceived his own quantum-mechanical theory not only to be in the spirit of Lewis' old idea of a paired-electron bond but to be a quantitative and more scientific version of it. Although Lewis was a scientist of the old school and a foreigner to quantum mechanics, he easily crossed the disciplinary boundaries between chemistry and physics. He was as much a chemical physicist and a physical chemist.

In the United States, the *Journal of Chemical Physics* was founded in 1933, in part as a substitute for and rival to the more traditionally oriented *Journal of Physical Chemistry* established in 1896 as an American counterpart to *Zeitschrift für Physikalische Chemie*. Contrary to traditional physical chemistry, the new chemical physics had a strong focus on quantum chemistry and molecular spectroscopy. In the first issue, Harold Urey — managing editor of the journal, the discoverer of deuterium, and a physicist as well as a chemist — reflected on the old theme of the relationship between physics and chemistry:

> "At present the boundary between the sciences of physics and chemistry has been completely bridged. Men who must be classified as physicists on the basis of training and of relations to departments or institutes of physics are working on the traditional problems of chemistry; and others who must be regarded as chemists on similar grounds are working in fields which must be regarded as physics" (Urey, 1933).

Although the valence bond and the molecular orbital methods appeared to be rival views of the chemical bond, a few scientists suggested early on that the two methods were complementary rather than antagonistic. According to John Slater, another of the pioneers of quantum chemistry and chemical physics, the choice between the two methods was largely a matter of convenience. This viewpoint became commonly accepted only after about 1950, at a time when the strength of the molecular orbital theory became increasingly recognized not least as a result of the important works of Charles Coulson, the eminent British theoretical chemist and applied mathematician. His pervasive arguments in favor of molecular orbitals were, to a large extent, responsible for the lack of popularity of the valence bond method. Coulson's book *Valence* first published in 1952 had an impact on the community of chemists of the same scale as Pauling's earlier *The Nature of the Chemical Bond*.

.2.5

Transuranic and Superheavy Elements

2.5

Transuranic and Superheavy Elements

We know for sure that there are no chemical elements lighter than hydrogen. Highly unstable atom-like systems such as positronium (Section 1.3.3) and muonium (where an electron orbits a positive muon) have atomic weights of ca. 0.01 and 0.11, respectively, and are thus definitely subhydrogenic. However, a chemical element is, by definition, given by its atomic number, and hence positronium and muonium do not qualify as proper elements. The question regarding the uppermost part of the periodic system is quite different, for still today we do not know the maximum number of elements. Presently element 118 ranks as the heaviest one, but it is possible and even likely that still heavier elements will be identified in the future.

The 26 known transuranic elements — heavier than uranium — have all been produced in the laboratory and thus not been discovered in the traditional sense of the word. They are unnatural, and some are more unnatural than others. Elements with atomic number larger than about 103 are often referred to as "superheavy elements." The history of the heavy artificial elements began in 1940, and element 118 was officially recognized as late as 2016. Thus, part of the history is not only modern history of science but contemporary history of science.

Whether natural or artificial, the discovery of new elements has always been regarded as something extraordinary and has, in many cases, been associated with controversies over priority. In a review of superheavy elements and their discoveries, the Dutch nuclear physicist Aaldert Wapstra wrote about the fascination of the discovery of new transuranic elements: "The problem is open although of final scope, unlike the number of continents upon the surface of the earth where we know with certainty that none still awaits discovery. These considerations give to the discovery of new elements an importance, an allure and a romance that does not attach to the discovery of, say, a new comet or a new beetle where many more such discoveries are to be anticipated in the future" (Wapstra *et al.*, 1991).

The synthesis and identification of transuranic elements, whether superheavy or not, rely crucially on advanced accelerator and detection technology and also on theoretical models of heavy atomic nuclei. In this respect, the field belongs to nuclear physics rather than chemistry, and yet chemical methods are no less significant than those of physics. For one thing, the identification of new nuclei relies on methods of what is called nuclear chemistry, a branch of the chemical sciences that emerged in the 1930s as a sub-discipline of the more traditional radiochemistry with roots thirty years earlier. For another and more important thing, since the eighteenth century matters concerning new elements have been considered as belonging to the domain of chemistry. The responsibility of recognizing new elements still belongs to International Union of Pure and Applied Chemistry (IUPAC) and not to the physicists' sister organization International Union of Pure and Applied Physics (IUPAP).

2.5.1. AN ERA OF SPECULATIONS

For a long period of time, the metallic element uranium with atomic number $Z = 92$ and atomic weight $A = 238$ was supposed to be the heaviest of the chemical elements. There might, in principle, be even heavier elements, but to almost all chemists this was a remote possibility that could safely be ignored. One of the few suggestions of transuranic elements, and possibly the earliest one, came from the Danish chemist Julius Thomsen, who in 1895, shortly after the discovery of argon, predicted a whole group of inert elements. The heaviest of these as yet hypothetical elements he assigned the atomic weight 292, but without suggesting that it actually existed. Nine years later, Charles Baskerville, a chemist at the University of North Carolina, claimed to have isolated a new element with $A = 255.8$ from thorium salts. "Carolinium," as he called it, failed to win recognition and was soon forgotten. As far as experience could tell, chemists and physicists unanimously agreed that uranium was at the very end of the periodic table.

But why uranium? Or, as the question was formulated after the recognition of the atomic number as the parameter of the periodic table, were there any reasons why atoms with a nuclear charge greater than 92 did not exist? If there were no compelling reasons, it made sense to theorize about atoms with more than 92 electrons revolving around the heavy nucleus. In the 1920s, questions of this kind were considered by a minority of physicists and chemists on the basis of the Bohr–Sommerfeld atomic model. Niels Bohr thought that the nuclei of atoms with $Z > 92$ would probably be highly radioactive and therefore unstable. Nonetheless, in his Nobel lecture of 1922 he presented the electron configuration of the inert gas with $Z = 118$, the same element which Thomsen had referred to 27 years earlier. Without believing in the actual existence of the element, Bohr suggested that the numbers of electrons arranged according to their principal quantum number were 2, 8, 18, 32, 32, 18, and 8.

The question of the maximum number of elements, or the end of the periodic table, was addressed by Bohr and some other quantum physicists. They argued that for a nucleus of very high Z, the innermost electron would not move around it in an elliptical orbit, but immediately fall into the nucleus. A simple calculation of the critical charge resulted in $Z_{max} = 137$ (the inverse of the fine-structure constant) and more sophisticated calculations suggested Z_{max} to lie in the neighborhood of 97. Theoretical estimates of this kind were not very convincing, and they did not rule out the existence of a least a small number of transuranic elements. On the other hand, they did not suggest that such elements really existed.

The German physicist and engineer Richard Swinne was the first to speak of "transuranic" elements, a name he introduced in a paper published in 1926. Based on theoretical as well as experimental arguments, Swinne went further than Bohr by suggesting electron configurations for all the elements in the interval $92 < Z < 108$. Moreover, he believed that traces of some of these elements might be found in iron meteorites or in the Earth's solid core. Swinne searched for the elements by means of X-ray spectroscopy, but his data were too uncertain to justify that they really existed. Other scientists also made use of Moseley's X-ray spectroscopic method to look for transuranic elements, and in 1934 there even was an unverified claim for element 93 in the form of the spurious "bohemium" announced by Odolen Koblic, a Czech chemical engineer.

The possibility of manufacturing new elements in the laboratory, rather than detecting them in nature, only became a realistic possibility after the discovery of the neutron in 1932. Initially, the neutron hypothesis was controversial, but within a year or so physicists accepted that the atomic nucleus consisted of protons and neutrons and not, as traditionally believed, of protons and electrons. As a neutron source, Enrico Fermi and his research group in Rome used alpha particles from radon or polonium colliding with beryllium nuclei: ${}^{4}_{2}\mathrm{He} + {}^{9}_{4}\mathrm{Be} \rightarrow {}^{12}_{6}\mathrm{C} + {}^{1}_{0}n$. In this way, they systematically studied neutron reactions with all elements, paying particular attention to the heavy elements. When they bombarded uranium with slow neutrons, they obtained results which suggested that the product was an element with atomic number greater than 92. Fermi and his collaborators believed they had produced elements 93 and 94, but for several years they cautiously avoided naming them, realizing that the interpretation of the experiments was after all uncertain.

When Fermi in 1938 was awarded the Nobel Prize in physics, it was in part for his "demonstration of new radioactive elements produced by neutron irradiation." In his presentation speech, the chairman of the Nobel Committee Henning Pleijel referred to the new elements as if they were now accepted members of the periodic system: "Fermi even succeeded in producing two new elements, 93 and 94 in rank number. These new elements he called Ausenium and Hesperium."[1] Fermi too referred to the elements in his Nobel lecture, but without claiming that he and his group had

[1] https://www.nobelprize.org/prizes/physics/1938/ceremony-speech/.

discovered them. According to Fermi, the two elements were possibly formed as decay products of the uranium isotope U-239:

$$^{238}_{92}\mathrm{U} + {}^{1}_{0}n \rightarrow {}^{239}_{92}\mathrm{U} \rightarrow {}^{239}_{93}\mathrm{Ao} + \beta^{-} \rightarrow {}^{239}_{94}\mathrm{Hs} + 2\beta^{-}.$$

Emilio Segré, one of Fermi's assistants, later recalled that Fermi initially refused to name the hypothetical elements, but that he, as a prominent scientist in Mussolini's Italy, was under some political pressure to do so:

> "I believe there were suggestions, if not outright pressures, to add glory to the fascist regime by giving to the hypothetical new elements some name reminiscent of the fascists. ... [Orso Maria] Corbino, who had an extremely prompt repartee, pointed out that the new elements had very short lives, and that this might make them inappropriate to celebrate fascism" (Segré, 1980, p. 205).

The lives of ausenium and hesperium were short indeed, as it was soon realized that they were mistakes. They went the same way as carolinium and bohemium — into the dustbin of history. If the incident was embarrassing for Fermi, it was no less embarrassing for the Nobel Committee.

There were other ways in which the idea of very heavy elements turned up in the interwar period. For example, it was sometimes suggested that to account for the presence of long-lived elements such as uranium and thorium, it was necessary to assume that they were decay products of even heavier radioactive elements. The highly respected physicist and astronomer James Jeans defended for almost a decade the unorthodox hypothesis of large amounts of transuranic elements in the stars and the nebulae. As he stated in a lecture of 1928, "The complete series of chemical elements contains elements of greater atomic weight than uranium, but all have, to the best of our knowledge, vanished from the earth, as uranium is also destined to do in time" (Jeans, 1928a). According to Jeans, only radioactive elements of $Z = 95$ or more could explain the generation of energy in nebulae and certain kinds of stars (see also Section 3.3.2). These elements were also present in the Sun, he believed, but only in its interior and thus were undetectable by means of spectroscopic means. Although the Earth was formed by

Solar matter, Jeans suggested that it mostly came from the Sun's surface.

Jeans presented his theory in several lectures and articles, and also in his monograph *Astronomy and Cosmogony* from 1928. It was thus well known, but the large majority of astronomers and physicists rejected it as speculative. As for the chemists, they ignored Jeans' theory. After all, there was not the slightest direct evidence for transuranic stellar elements. From about 1925 to 1938, the German physical chemist and Nobel Prize laureate Walther Nernst held views somewhat similar to Jeans' if motivated differently. Nernst and Jeans both conjectured that cosmic matter originally consisted of complex unstable elements of very high atomic weight and that these eventually gave rise to the presently observed elements. This unorthodox view went counter to the generally accepted view of element formation taking place by synthesis of light elements and not by disintegration. Contrary to Jeans, Nernst believed that elements much heavier than uranium might still exist in the crust of the Earth. When he got news of Fermi's experiments with artificially produced transuranic elements, he took it as a vindication of his ideas.

Without referring to Bohr, Jeans, or Nernst, in 1928 an American scientist at the General Electric Company speculated in a paper in *Scientific Monthly* that stellar energy had its origin in a hypothetical element of atomic number 118. W. S. Andrews, as his name was, thought that the regularity of the periodic table justified his speculation of the supposedly super-radioactive element, and contrary to earlier speculators he gave it a name, if not a chemical symbol. He called it "hypon."

2.5.2. TOWARDS SUPERHEAVY ELEMENTS

Whereas Fermi's announcement of ausenium and hesperium was premature, the announcement of neptunium and plutonium about two years later was real and of great importance. The discoveries were closely connected to the sensational discovery of the fission of the uranium nucleus still unknown when Fermi gave his Nobel lecture in Stockholm. Investigations of fission fragments led the Berkeley physicists Edwin McMillan and Philip Abelson to conclude that they had detected element 93 (neptunium, Np) with atomic weight 239 as a decay product of the isotope U-239. This first transuranic element was announced without much fanfare in *Physical Review* on 27 May 1940. A few weeks later, *Science News Letter* covered the discovery under the misleading title "Superheavy Element 94 Discovered in New Research," suggesting that McMillan and Abelson had "confirmed the discovery made several years ago by Prof. Enrico Fermi" (Anon, 1940).

Shortly later, Glenn Seaborg and collaborators, also at Berkeley, prepared the Np-238 isotope by bombarding uranium with deuterons accelerated in a cyclotron:

$$^{238}_{92}\mathrm{U} + ^{2}_{1}\mathrm{H} \rightarrow ^{238}_{93}\mathrm{Np} + 2\,^{1}_{0}n.$$

The Np-238 isotope, they proved, decayed into an isotope of element 94, which came to be known as plutonium:

$$^{238}_{93}\mathrm{Np} \rightarrow ^{238}_{94}\mathrm{Pu} + \beta^{-}.$$

Importantly, the Berkeley group also produced the isotope Pu-239, which during the war was produced in large quantities and used in the nuclear bomb that destroyed Nagasaki on 9 August 1945. Although the discovery paper of Seaborg and his group was submitted on 7 March 1941, because of the war it only appeared in print more than five years later and it was also only then that the name "plutonium" became generally known.

Among the many transuranic elements, the silvery-white metal plutonium has a unique status which is not exclusively related to its political and military significance. It is estimated that today the world stockpile of the element is about 400 tons corresponding to about 10^{40} atoms. The long half-lives of the plutonium isotopes Pu-239 (2.4×10^4 years) and Pu-244 (8×10^7 years) means that the element is not just an ephemeral visitor on Earth but will remain with us for thousands of years to come. Traces of plutonium have been found in nature, but in extremely small amounts and always in connection with uranium deposits. The plutonium from these sites is the result of rare nuclear processes in relatively concentrated uranium, whereas other traces of plutonium found in nature are due to the nuclear tests carried out in the 1950s and 1960s.

In 1971, Darleane Hoffman and coworkers reported that they had found traces of primordial Pu-244 in old rocks, suggesting that the 2.4×10^4 detected atoms were left over from the formation of the solar system nearly five billion years ago. However, later experiments have not confirmed the existence of primordial plutonium on Earth, and today the results from 1971 are no longer believed.

Still during the war, the elements 95 (americium, Am) and 96 (curium, Cm) became realities. While the first was identified as a beta decay product of Pu-241, the latter was produced by bombarding Pu-239 with a beam of alpha particles:

$$^{239}_{94}\mathrm{Pu} + {}^{4}_{2}\mathrm{He} \rightarrow {}^{242}_{96}\mathrm{Cm} + {}^{1}_{0}n.$$

When IUPAC reviewed the state-of-art at a conference in 1949, the four transuranic elements and their names were officially recognized. This was only the beginning of what soon became a competitive race to synthesize still heavier elements.

The further development was primarily experimental, but it also stimulated interest from nuclear theorists trying to understand the increasingly bigger and complex atomic nuclei. One of the theorists was the American John Wheeler, a former collaborator of Bohr and a specialist not only in nuclear physics but also in general relativity theory. In the 1950s, Wheeler examined, theoretically, the limits of nuclear stability and in this context suggested the name "superheavy element." The name, frequently abbreviated SHE, was adopted by later researchers and is today generally used although with somewhat different meanings. A superheavy element is often taken to refer to elements with atomic numbers ranging from 104 to 121, the so-called transactinide series, but there is no sharp distinction between the terms transuranic and superheavy.

According to the shell model of the atomic nucleus developed in the late 1940s, nuclei with 2, 8, 20, 50, 82, and 126 protons or neutrons were particularly stable. The numbers were "magical," representing closed shells in the nucleus. Later theories based on the shell model indicated that the number 114 was also magical and that the nucleus $(A, Z) = (298, 114)$ might be accessible for laboratory studies. Considerations of this kind led in the 1960s to the idea of an "island of stability" represented by nuclei with a relatively long half-life. A favored center of the hypothetical region was the nucleus with 114 protons and 184 neutrons, but there were other possibilities. According to the Swedish physicist Sven Gösta Nilsson, the half-lives of nuclei in the island of stability, such as $(A, Z) = (294, 110)$, might be as long as 100 million years. If so, there might still be traces of them. The hope of the experimenters was to reach the fabled island, if it existed, either by manufacturing the elements or by finding them in nature. The first method was far more expensive than the second, but it was also the more realistic one.

A German nuclear chemist and specialist in SHEs wrote about the impact of the island of stability: "These predictions of very long half-lives immediately stirred up a gold-rush period of hunting for superheavy elements in natural samples. Everybody was encouraged to participate. ... Just an intelligent choice of a natural sample and a corner in the kitchen at home could be sufficient to make an outstanding discovery: new and superheavy elements in Nature" (Hermann, 2014). Indeed, from about 1970 many chemists and physicists, and a few amateur scientists as well, began looking for evidence in cosmic rays, meteorites, terrestrial ores, geothermal waters, and other sources. Even samples of lunar matter were examined. The preferred method was to look for tracks due to spontaneous fission, a radioactive process that is exceedingly rare in nature but is the dominant decay mode for superheavy elements. Although spontaneous fission in uranium was identified as early as 1940, the probability that a U-238 nucleus splits spontaneously into two fragments is negligible.

Hundreds of searches for natural SHEs have been conducted, and in several cases positive results were announced only to be retracted shortly after. In 1973, a team of American scientists conducted an extensive investigation of SHEs in terrestrial and meteoric minerals, but the result of their search was negative. Three years later, another American team concluded optimistically to have found convincing evidence for the three elements 116, 124, and 126. The announcement caused excitement not only in the physics community but also in the mass media. Under the headline "Discovery of superheavy elements" *Nature* reported, prematurely as it turned out, that now the three elements had been detected and found to be stable. Alas, when the confirming evidence was critically examined by other specialists, the discovery claim lost its credibility. Also, Russian scientists searched eagerly for natural SHEs, and on some occasions they too came up with premature announcements of success.

While most SHE hunters focused on terrestrial and atmospheric sites, others examined the possibility that SHEs were produced in stellar nuclear reactions. This was a plausible hypothesis given the history of the element technetium with $Z = 43$, which was the first artificially produced element ever. In 1937, it was identified in nuclear reactions by Segré and his collaborator Carlo Perrier, and for more than a decade it was believed to be absent in nature. Only in 1952, when absorption lines due to technetium were observed in the spectra of giant stars, did it turn out that the element is formed in natural processes. Perhaps something similar might be the case for one or more of the superheavy elements. But again, astronomical data provided no evidence that SHEs play a role in stellar nuclear reactions. By the mid-1980s, practically all experts agreed that SHEs belong to the laboratory and not to nature. Although this is still the consensus view, the search for natural SHEs has continued until the present.

2.5.3. NOBELIUM, ELEMENT 102

Element 101, named mendelevium and assigned chemical symbol Md, was one more triumph of the Berkeley group and their use of alpha particles as projectiles in nuclear synthesis. The target of the group's experiments in 1955 was a tiny sample of the element einsteinium discovered a few years earlier:

$${}^{253}_{99}\mathrm{Es} + {}^{4}_{2}\mathrm{He} \rightarrow {}^{256}_{101}\mathrm{Md} + {}^{1}_{0}n.$$

Only 17 atoms of Md-256 were identified, but this was enough to establish the discovery. Up to that time, the synthesis of new transuranic elements was completely dominated by Seaborg and his group of Californian physicists and nuclear chemists. All the elements from 93 to 101 were produced in reactions with light projectiles, either deuterons, neutrons, or alpha particles. In the next generation of transuranic elements, starting with atomic number 102, the projectiles were more heavy ions, such as boron, carbon and oxygen. Moreover, the Americans were no longer the only party in the game, as they began to face stiff competition from the Joint Institute for Nuclear Research (JINR) founded in Dubna outside Moscow in 1956. A third party only entered around 1970 with the establishment of an important center of SHE research in Darmstadt, West Germany, and a fourth party was added some twenty years later in the form of a Japanese institution called RNC or the RIKEN Nishina Center.

The question of priority to the discovery of element 102 involved a major controversy between the Americans and the Russians, but it started with a third group of contenders. In 1957, an international team of four Swedes, two Britons, and one American announced that they had detected four atoms of element 102 by bombarding a sample of curium with C^{4+} ions of mass number 13. Since the work took place at Stockholm's Nobel Institute of Physics, the group suggested to call the new element nobelium in honor of Alfred Nobel. From measurements of the secondary alpha particles emitted in the process, the group believed it to be

$$^{244}_{96}\mathrm{Cm} + {}^{13}_{6}\mathrm{C} \rightarrow {}^{253}_{102}\mathrm{No} + 4\,{}^{1}_{0}n,$$

or possibly resulting in No-251 and six neutrons. The discovery claim created much attention in Swedish and British news media, not least because it was the first transuranic element discovered in Europe. However, the Stockholm claim turned out to be unfounded or at least insufficient. It was only the beginning of a much longer and more complex controversy with American and Russian scientists as the main competitors.

The results obtained in Stockholm could not be reproduced in Berkeley, where a team led by Albert Ghiorso and including Seaborg in 1958 claimed to have proved the existence of element 102. The experiments of the Berkeley group made use of a new heavy-ion linear accelerator called HILAC, but the reactants were largely the same as in the Stockholm experiments except that the Americans used C-12 ions as projectiles and Cm-246 as target material. Faced with the American claim, the Stockholm group admitted that the evidence from 1957 was somewhat ambiguous and that the original discovery claim was in need of further support. The results of the Stockholm group could also not be reproduced by Georgii Flerov and his group in Dubna, which initially experimented with O-16 ions and Pu-241 as target. Nor did the Russians accept the Berkeley claim of priority based on the 1958 experiments.

After an extended series of further experiments, the Russians concluded that they had identified element 102 in the reaction

$$^{238}_{92}\text{U} + ^{22}_{10}\text{Ne} \rightarrow ^{256}_{102}\text{No} + 4\,^{1}_{0}n.$$

In a later report, the Dubna group summarized: "Element 102 was discovered at Dubna in studies carried out during 1963–1966. Those studies contain unambiguous and exhaustive proof of the synthesis and observation of nuclei of this element, and correct data of the characteristics of the α decay properties of five isotopes of element 102" (Flerov *et al.*, 1991). What had at first been a controversy between three parties became a Russian–American controversy fueled by the atmosphere of mutual disbelief characteristic of the Cold War era and surviving after it ended. Ghiorso, Seaborg, and their colleagues responded by maintaining the validity of the work done at Berkeley and criticizing the Russian results, which they did not accept as "unambiguous and exhaustive proof" at all.

It took decades before the conflict over name and priority of element 102 was finally settled. As far as the name is concerned, "nobelium" had quickly come into common usage and entered textbooks and periodic tables. At a conference in 1961 — at a time when it was known that the Stockholm claim was more than questionable and the existence of the element was still contested — IUPAC decided to approve the name and the symbol No. The Americans were not happy but refrained from proposing an alternative name. As they wrote in a paper of 1967: "In view of the passage of time and the degree of use of this honorable name in numerous text books and scientific writings throughout the world, we wish to suggest that 'nobelium' be retained as the name along with its symbol, 'No'" (Ghiorso and Sikkeland, 1967). Nor did the Russians protest, but for a period of time they expressed their dissatisfaction by avoiding the name sanctioned by IUPAC. Instead they preferred to speak in their publications of just element 102 or sometimes of "joliotium." The latter name was a reference to the

French radiochemist and nuclear physicist Frédéric Joliot-Curie, who together with his wife Irène Joliot-Curie had discovered artificial radioactivity and for this work were awarded the Nobel Prize. Frédéric Joliot-Curie was also a devoted communist and a staunch supporter of the Soviet Union, which presumably contributed to the Russians' preference for the name. Whereas the Americans never considered joliotium, in 1992 the German Darmstadt group found the name acceptable as a compromise in the naming controversy.

To take care of the many priority disputes, in 1985 IUPAC and its sister organization IUPAP established a joint Transfermium Working Group (TWG) consisting of seven nuclear physicists and two nuclear chemists. Initially, the working group consisted solely of physicists, and it was only after objections from the chemistry community that it was extended with two IUPAC representatives. To avoid national bias, scientists from the Soviet Union, USA, and West Germany, the three countries with SHE laboratories, were excluded from the TWG panel. After a review of all relevant papers on element 102, in a report of 1992 the TWG concluded in favor of the Dubna team rather than the competing Berkeley team (the claim of the Stockholm team had long been disregarded). The decision, which was approved by IUPAC, caused vehement reactions from the Americans, who charged that the TWG panel was incompetent and biased. They felt in particular that the panel was too much physics-focused, whereas it downgraded the contributions of nuclear chemistry which were central to the American claim.

The indignant protests of the Berkeley scientists were heard by IUPAC but without causing the organization to change its decision. This was the end of the story as far as priority was concerned, but the question of the name came up once more. In 1995, as part of a renewed discussion of the names of the very heavy elements, IUPAC proposed to replace nobelium with "flerovium" (symbol Fl) in commemoration of Flerov, who had died five years earlier. However, the proposal did not win official recognition, and in 1997 nobelium was restored as the one and only name for element 102. The name flerovium returned on the scene in 2012, when it was instead assigned for a new element with atomic number 114 discovered in 1998.

2.5.4. TRANSFERMIUM WARS

The disputes concerning the discovery of element 102 were followed by disputes of a similar nature related to the discovery claims of elements with atomic numbers ranging from 103 to 109. These elements were all "transfermium," meaning beyond the element fermium with $Z = 100$. They were synthesized in the two decades from 1965 to 1985 in a competitive race between scientists from Dubna and Berkeley, and since the mid-1970s also with participation of scientists from the Darmstadt group. The method was in all cases heavy-ion fusion reactions where a target of a heavy element was bombarded with medium-sized ions. For example, element 104 was first claimed detected in Dubna in 1964 by bombarding Pu-242 with Ne-22 ions. The Russian evidence for the new element was contested by the Berkeley group who, five years later, announced to have found what was missing, namely undeniable evidence for the element produced in the reaction

$$^{249}_{98}\mathrm{Cf} + {}^{12}_{6}\mathrm{C} \rightarrow {}^{257}_{104}\mathrm{Rf} + 4\,{}^{1}_{0}n.$$

Unimpressed, Flerov and his group at Dubna maintained their priority claim to what the Americans called rutherfordium (Rf) but the Russians kurchatovium (Ku) in honor of the Soviet physicist Igor Kurchatov. It took until 1997 before IUPAC finally decided in favor of the present name rutherfordium. The alternative name kurchatovium appeared regularly in scientific journals from the Soviet Union and its allied countries, but to scientists in the West it was anathema. Kurchatov was director of the Soviet nuclear weapons program and a main contributor to the country's hydrogen bomb, and for this reason alone it was unacceptable to name an element after him.

The priority controversy over element 104 was far from unique, for similar controversies took place also in the case of other of the transfermium elements. One of the more problematic elements was number 106, seaborgium (Sg), where the priority controversy was mixed up with a protracted dispute over its name, which was proposed by Ghiorso and his team in 1994. Seaborg, the Nobel laureate and distinguished veteran in transuranic science, was at the time 82 years old, and it had never happened before that an element had been named after a living scientist (einsteinium was close, but the name was approved by IUPAC only in 1957, two years after Einstein's death). After much discussion, in 1997 IUPAC finally accepted seaborgium as the name of element 106. Seaborg could thus enjoy having an element named after him for the two years remaining of his life.

The series of controversies clearly indicated that there was a need to establish common and operational criteria for what it means to have discovered a new element beyond the shared criterion of the atomic number as the defining parameter. The criteria had to be acceptable to both physicists and chemists, and to meet the approval of all research groups engaged in element synthesis. As it turned out, this was no easy matter. The situation eventually escalated to what the American nuclear chemist Paul Karol in 1994 called the "transfermium wars," a series of convoluted disputes concerning the names and discoveries of the new synthetic elements.

On the proposal of the Dubna group, in 1974 IUPAC and IUPAP appointed a joint working group of nine neutral experts, three of whom were from the United States and three from Soviet Russia. The responsibility of the group was limited to the Berkeley–Dubna rivalry concerning elements 104 and 105, but this first initiative was a complete failure. In fact, the committee never met as a group and it never issued a report. It was only with the establishment in 1985 of the TWG that something happened. The new group headed by Denys Wilkinson, a distinguished Oxford nuclear physicist, had two main responsibilities. One was to formulate criteria for when an element was discovered, and the other was to evaluate discovery claims in accordance with these criteria. On the other hand, proposals of names were not the business of Wilkinson's group, but of an IUPAC body called the Commission on Nomenclature of Inorganic Chemistry or just CNIC. The TWG took both responsibilities seriously such as shown by two comprehensive reports published in 1991 and 1992 in the IUPAC journal *Pure and Applied Chemistry*.

In the first report, the TWG dealt systematically with the criteria that must be satisfied for a discovery claim of a new element to be recognized as a discovery. The group realized the complexity of the concept of discovery and the legitimacy of dissenting views of when and by whom an element was discovered. In the end, the concept was a consensus convention and could therefore be criticized or change over time. For example, when the concept of an element changed in the early 1920s, so did the meaning of having discovered a new element (Section 2.4.2). The TWG's best offer of a summary definition was this: "Discovery of a chemical element is the experimental demonstration, beyond reasonable doubt, of the existence of a nuclide with an atomic number Z not identified before, existing for at least 10^{-14} s" (Wapstra *et al.*, 1991). The requirement of a minimum lifetime of the identified nucleus was introduced to make the definition agree with the standard chemical view of an element as consisting of atoms. In most SHE experiments, nuclei and not atoms were identified, and it takes about 10^{-14} s for a heavy nucleus to acquire its electron system and form an atom. While atoms can be assigned chemical properties, nuclei cannot.

The phrase "beyond reasonable doubt" in the TWG definition may appear to be intolerably vague, but the group used it deliberately to stress that such doubts could not be completely avoided. As the TWG report phrased it, "the situation in respect of the discovery of a new element is by no means always black-and-white in the sense that it may be unequivocally asserted that a new element was discovered, with the required certainty, by a certain group, using a certain method on a certain date." Independent confirmation of a discovery claim was of course important, but it was not a magic wand that would in all cases eliminate doubts. In fact, TWG suggested that in some cases a discovery claim could be accepted even without a repetition of the experiment. Whereas the careful formulations in the 1991 report were generally accepted by the three SHE parties, this was not the case with the 1992 report assigning priority and credit for the discoveries of elements with atomic numbers from 101 to 112. Unsurprisingly, the conclusions of this report were received in widely different ways by the three involved research groups.

While in some cases the TWG conclusions with respect to priority were unambiguous, in other cases they were not. Thus the TWG group, after having reviewed all papers on element 103, lawrencium (Lr), suggested that credit for having discovered the element should be shared between Berkeley and Dubna. IUPAC followed the recommendation of this discovery being a co-discovery. TWG similarly suggested American–Russian co-discovery in the cases of $Z = 104$ and $Z = 105$. The three SHE laboratories were given the right to respond to the 1992 TWG report, which they promptly did. Whereas the Russian and German scientists were reasonably satisfied, the Americans were frustrated and dissatisfied. Seaborg and Ghiorso openly suggested that in the case of element 104, the TWG had applied a double standard in the evaluation of the work by the Dubna and the Berkeley groups. The two Americans denied that Dubna had a legitimate share in the discovery of the element, and they charged that the TWG group, far from being impartial, was in reality pro-Dubna and anti-Berkeley.

The protest had no effect as Wilkinson and his colleagues decided that there were no good reasons to change the conclusions of the TWG report. In a report from 1993, after the TWG had been terminated, he indirectly responded to the charges of Seaborg and Ghiorso. Referring to the thousands of hours that the working group had spent in a "microscopic and scrupulous analysis of the discovery of the transfermium elements," Wilkinson emphasized the objectivity of TWG: "We were utterly without bias, prejudice or pre-commitment and had no connection with any of the laboratories of chief concern; we did not care who had discovered the elements in question but agreed to find out" (Wilkinson *et al.*, 1993). In the series of controversies related to the discoveries of transfermium elements, some of the involved nuclear chemists complained that the TWG panel was dominated by physicists who did not fully appreciate the methods of chemistry (see also Chapter 2.4). Much later, the Swedish theoretical physicist Cecilie Jarlskog, a former member of the Nobel Prize physics committee, turned the table around. In an address in 2016, she accused the chemical community in the form of IUPAC to have unfairly stolen the credit from the physicists.

2.5.5. THE HEAVIEST ELEMENT SO FAR

The brief period of the transfermium war did not prevent further progress in the manufacture of still heavier elements. By the mid-1990s, the Americans and Russians were beginning to collaborate in a series of experiments made in Dubna. The teams headed by the Armenian-Russian SHE veteran Yuri Oganessian included scientists from the Lawrence Livermore National Laboratory (LLNL) and also from the Oak Ridge National Laboratory (ORNL) in Tennessee. In 2016, four newcomers to the periodic table were officially recognized by IUPAC, namely the elements with atomic numbers 113 (nihonium, Nh), 115 (moscovium, Mc), 117 (tennessine, Ts), and 118 (oganesson, Og). Three of the four elements were products of the successful Russian–American collaborations. The fourth, nihonium, also deserves attention since it was credited the RIKEN research center in Japan and the first element ever to be discovered in Asia by scientists from this continent. Kōsuke Morita and his collaborators achieved their first result by bombarding a Bi-209 target with Zn-70 projectiles for nearly a full year. As they reported in 2004, they were able to identify a single atom of the new element with mass number 278.

IUPAC's approval of element 118 was primarily based on Dubna experiments made in 2006 by Oganessian's large team of 20 Russian scientists and 10 Americans from the LLNL. The team observed three chains of alpha decays arising from the fusion of Cf-249 with Ca-48 ions, which they interpreted as

$$^{249}_{98}\mathrm{Cf} + ^{48}_{20}\mathrm{Ca} \rightarrow ^{294}_{118}\mathrm{Og} + 3\,^{1}_{0}n.$$

Of course, at the time the still unrecognized element 118 was not yet named oganesson with the symbol Og. The name accepted by IUPAC 10 years later was in honor of the then 83-year-old Oganessian and is presently the only name for an element referring to a living scientist. When "oganesson" was chosen instead of "oganessium" it was to underline that the element belongs to group 18 in the periodic table, such as do xenon, radon, and other noble gases. The only exception to the suffix -on is helium, which sensibly has not been renamed "helion," although this was what a British chemist proposed in 1927.

Like several other superheavy elements, the one with atomic number 118 had an earlier history. As a purely hypothetical element, it appeared as early as 1895, and in the 1920s Bohr and Swinne suggested electron structures for it (see Section 2.5.1). In 1965, a German-American radiochemist by the name Aristid

Grosse made detailed calculations of the element's physical and chemical properties. Adopting a nomenclature used by Mendeleev, Gross referred to the element as "eka-emanation," where "emanation" is an older name for radon. Of more relevance, in the summer of 1999 the journal *Science* could tell its readers that "Berkeley group bags element 118." The heading referred to a recent paper in *Physical Review Letters* in which a group of researchers at Lawrence Berkeley National Laboratory (LBNL) reported evidence for three cases of the fusion process

$$^{208}_{82}\mathrm{Pb} + {}^{86}_{36}\mathrm{Kr} \rightarrow {}^{293}_{118}\mathrm{Og} + {}^{1}_{0}n.$$

The lead author of the paper was 39-years-old Viktor Ninov, an experienced Bulgarian–German physicist who had previously worked at the Darmstadt laboratory and at LBNL took care of the computer program analyzing the experimental data.

The news from Berkeley were exciting and at first generally accepted, but neither the Darmstadt scientists in Germany nor the RIKEN scientists in Japan were able to confirm the data. Even worse, element 118 failed to turn up in repetitions of the experiment made at LBNL the following year. What had gone wrong? A formal investigation committee concluded in 2002 that some of the crucial data were fabricated and that Ninov was guilty in the fabrication. Although Ninov indignantly denied the charges, in May 2003 he was fired from LBNL and the rest of the group hurriedly retracted the claim for the synthesis of element 118. The embarrassing case of scientific misconduct attracted wide public attention and also raised the question of the responsibility of the other authors of the 1999 discovery paper. Moreover, how could this kind of outright fraud be avoided in the future? Most SHE scientists just wanted to forget about the unfortunate incident. Ninov ended up not only as a *persona non grata*, but as a person who had never existed. For example, in their discovery paper of 2006, Oganessian and his coauthors only referred *en passant* to the experiment of 1999 (which "was later disproved") and without referring to Ninov or the retracted paper.

Although the discovery of element 118 claimed by Ninov and the other members of the LBNL group was definitely wrong, after 2016 the element was declared real. But is it? Or rather, in which sense is the element real? These are not silly questions, for the superheavy elements are in an ontological sense very different from other chemical elements such as oxygen, iron, and uranium, or even the transuranic americium widely used in smoke detectors. What is "known" about the physical and chemical properties of oganesson are entirely the result of predictions and extrapolations. For example, contrary to earlier assumptions the element is believed to be a reactive solid with a density of ca. $5 g/cm^3$ and a melting point of ca. 50°C. There are no empirical data and none are expected in the foreseeable future. For element 117, tennessine, the situation is the same.

The half-lives of the longest-lived transfermium isotopes vary greatly and generally decrease with the atomic number, from 51.5 days for Md-258 to 0.69 ms for Og-294, the oganesson isotope detected in 2006. This implies that, in several cases, the elements have been produced and verified only to disappear again almost instantly. As noted by two Darmstadt scientists involved in the discovery of element 109 (meitnerium, Mt): "Once synthesized, elements such as 109 decay so rapidly that synthesis cannot keep up with decay. The heavier elements are so short-lived that by the end of the irradiation all atoms created have already decayed. These atoms must therefore be detected and identified during the production process itself" (Armbruster and Münzenberg, 1989).

So, strictly speaking the heaviest of the transuranic elements do only *exist* in the abstract sense that the *Tyrannosaurus rex* or the Tasmanian tiger exists. These animals of the past once *existed* but are now extinct. Likewise, the superheavy elements once existed, namely in a brief moment of the discovery experiments, but they no longer do so. Can one reasonably say that the element oganesson exists when there is not, in all likelihood, a single atom of it in the entire universe? More atoms or rather nuclei of the element could be produced by repeating or modifying the Dubna experiments, but within a fraction of a second the new oganesson nuclei would disappear again. Such is the strange new world of modern SHE research.

Part 3

Astronomy and Astrophysics

3.1

Curved Space:

From Mathematics to Astronomy

3.1

Curved Space: From Mathematics to Astronomy

Little is known about the life of Euclid, the Greek mathematician who, in about 300 BC, wrote a treatise called the *Elements*, possibly the most important and easily the most influential work in the history of mathematics. He may have worked in the famous library of Alexandria shortly after it was founded by Alexander the Great, but even this is uncertain. Euclid founded plane geometry on five postulates or axioms, and from these and a few definitions he deduced logically a number of well-known theorems, such that the circumference of a circle with radius R is exactly $2\pi R$ and that the angle sum of any triangle equals 180°. The basic axioms were independent and in no need of justification, but the fifth of them, the parallel postulate, was different from and less obvious than the other ones. According to this postulate, two parallel lines never intersect; or more precisely, through a point not on a given straight line, at most one line can be drawn that never meets the given line.

Many mathematicians following in the footsteps of Euclid wondered if the parallel postulate was really a necessary truth. If this were not the case, a new kind of geometry was a logical possibility, a hypothetical space where parallel lines do intersect and where the circumference of a circle differs from $2\pi R$. In about 1830, mathematicians proved the existence of non-Euclidean geometries corresponding to so-called curved spaces. The new geometries were consistent and in this sense existed mathematically, but did they also describe the real space in which physical and astronomical events unfold? Although non-Euclidean geometry was much discussed during the last third of the nineteenth century, few scientists believed that it had anything to do with physical space or that measurements could decide whether space is Euclidean or not.

The geometrical theorems derived in Euclid's *Elements* were two-dimensional, but the Greeks knew of course that real space has three dimensions. No more and no less. During the period when non-Euclidean geometry was first discussed, the strange idea of a fourth space dimension was also introduced. The two concepts are independent, and yet they were often grouped together in what was called meta-geometry or sometimes transcendental geometry. In both cases, twentieth-century physics turned the speculations into proper scientific theories. Today, we know that space is generally curved and hence non-Euclidean, whereas the existence of more than three space dimensions still belongs to the realm of theory.

3.1.1. NON-EUCLIDEAN GEOMETRY

According to the famous enlightenment philosopher Immanuel Kant, Euclidean space was not an option but a necessity for rational thinking about the world. Since it was true a priori and a presupposition for knowledge, it could not possibly be wrong. But even the greatest philosophers may be mistaken, as Kant was in this regard and as was first demonstrated by Karl Friedrich Gauss, a professor at the University of Göttingen in Germany.

Gauss was a true genius, one of the greatest mathematicians ever. His contributions to science were not limited to pure and applied mathematics, though, for he was also a pioneer in celestial mechanics, electrodynamics, terrestrial magnetism, and much more. In about 1815, Gauss reached the conclusion that Euclid's venerable geometry is not true by necessity but can only be justified by means of precise observations. Geometry, he thought, belongs to the empirical sciences and hence must be judged on the basis of measurements. Gauss communicated his startling insight in letters to colleagues in mathematics and astronomy, but without elaborating or writing it down in a paper. While Gauss anticipated geometries different from Euclid's flat-space geometry, it was left to the Hungarian János Bolyai and the Russian Nikolai Ivanovich Lobachevsky to establish non-Euclidean geometry as a new branch of mathematics.

Of the two pioneers, Lobachevsky is the more interesting since he clearly contemplated the problem within an astronomical perspective. Lobachevsky suspected, much like Gauss did, that the truth of geometry, Euclidean or not, "can only be verified, like all other laws of nature, by experiment, such as astronomical observations" (Lobachevsky, 1898, p. 67). As was shown around 1870, there are three and only three possible geometries, each characterized by a space curvature K and its associated radius of curvature R. The relation between the two quantities is given by a particular value of the dimensionless curvature constant k, namely as

$$R^2 = \frac{k}{K}.$$

Thus, the curvature K has the dimension of an inverse area and the greater it is, the smaller is the curvature radius R. The curvature constant can attain three values, with $k = 0$ for the flat or Euclidean space; $k = +1$ corresponds to a spherical space and $k = -1$ to a hyperbolic space. To illustrate the difference between the three cases, consider again the sum of angles in a triangle. The well-known result 180° holds true for flat space, whereas the sum of angles is greater in a spherical triangle and smaller in the hyperbolic case.

Lobachevsky, a professor of mathematics at the University of Kazan some 800 km east of Moscow, was also trained in astronomy and had served as director of the university observatory. In a Russian treatise from 1829, he suggested that astronomical space might possibly satisfy what he called an "imaginary" geometry. This kind of non-Euclidean geometry soon came to be known as hyperbolic rather than imaginary. Lobachevsky argued that his hypothesis of triangles with an angle sum less than 180° might be checked by using the heavenly triangle consisting of the Sun, the Earth, and the star Sirius. However, available data resulted in a deviation from the Euclidean value of 180°, which was so tiny that it could easily be accommodated by the error of measurement. Lobachevsky cited a deviation of only 0.00072″ (arc seconds) or essentially zero. In other words, observational reasons seemed to confirm that space was Euclidean.

Nonetheless, this was not what Lobachevsky concluded. He realized that although measurements could, in principle, prove space to be curved or non-Euclidean, because of the unavoidable observational errors they could never prove space to be exactly flat or Euclidean. For this reason, he regarded his calculations to be inconclusive without seriously disbelieving the Euclidean nature of space. In a later publication, Lobachevsky considered a triangle spanned by a star and the two positions of the Earth half a year apart in its orbit around the Sun. The angle at the star α is twice the value of what is known as the stellar parallax p. Denoting the two angles at the positions of the Earth as β and γ, the general relationship between the angles is

$$p = \pi - (\beta + \gamma) = \alpha - K\beta.$$

In the Euclidean case $K = 0$, the parallax equals α and tends toward zero as the distance increases toward infinity, but this is not the case for a hyperbolic or spherical space. Lobachevsky realized that in hyperbolic space there is a minimum non-zero parallax for all stars irrespective of their distances from the Earth. In principle if not in practice, this provided a test of the geometry of space. Lobachevsky also applied his new geometry to the universe at large, suggesting that, if cosmic space is hyperbolic, the imaginary curvature radius R must be greater than 3×10^5 AU (astronomical units) or 5 light years. However, at the time nothing was known about the size of the universe.

Neither Gauss, Bolyai, nor Lobachevsky recognized the possibility of positively curved space given by $R^2 = 1/K$. This important class of non-Euclidean geometry was introduced by the eminent German mathematician and physicist Bernhard Riemann in a famous address in 1854. As Riemann realized, if space is characterized by the curvature constant $k = +1$, it must necessarily be closed and therefore of finite extent, and yet it has no boundaries. Intuitively we associate unboundedness with infinity, but this is only because our mind is Euclidean. Consider the surface of a spherical body with radius R, such as our globe with $R = 6371$ km. The surface has a finite area given by $4\pi R^2$, but it has no boundary — we can walk in any direction for an infinity of time without leaving the surface. Spherical space is three-dimensional and not two-dimensional as the sphere's surface, so it is just an analogy where we have suppressed one of the space dimensions.

The spherical universe is difficult to visualize and easily confused with a sphere in ordinary flat space. In the latter case, the volume of the sphere is $4/3\pi R^3$, where the radius R is the constant distance from the surface to the center of the sphere. Spherical space is quite different. Although it has a volume given by the curvature radius, namely $V = 2\pi^2 R^3$, it has no more a center than the two-dimensional surface of a Euclidean sphere has. Is the Earth at the center of the universe? Or is it the Sun? The closed spherical universe avoids this kind of futile questions by declaring a cosmic center an illegitimate idea.

Riemann expressed his insight as follows: "In the extension of space-construction to the infinitely great, we must distinguish between *unboundedness* and *infinite extent*. ... If we ... ascribe to space constant [and positive] curvature, it must necessarily be finite provided this curvature has ever so small a positive value" (Riemann, 1873). However, Riemann's comments of 1854 were sparse and enigmatic, and he was not seriously interested in the space of the astronomers. As far as he was concerned, questions about the global properties of the space of the stellar universe were metaphysical and could safely be left to the philosophers. In any case, his brilliant address of 1854 was only published *post mortem*, in 1867, and for this reason its impact was much delayed. Non-Euclidean geometry circulated at first slowly in the mathematical community, but since the mid-1870s the ideas of Lobachevsky, Riemann, and others became well known and attracted much interest, not only among mathematicians but also among philosophers and the general public. In the period from 1870 to 1910, more than 4000 titles were published on the subject. On the other hand, only a dozen physicists and astronomers felt tempted to investigate the remote possibility that real space should deviate from Euclid's flat geometry.

The extent to which the counterintuitive ideas of non-Euclidean space permeated the cultural and literary segment of society may be illustrated by the famous Russian author Fyodor Dostoevsky's classic novel *The Brothers Karamazov* published 1879–1880. In one of the passages, Ivan Karamazov confides to his younger brother Alyosha that he, although a believer, does not understand the nature of God any better than he understands the new geometry:

> "If God exists and if He really did create the world, then, as we all know, He created it according to the geometry of Euclid. ... Yet there have been and still are geometricians and philosophers, and even some of the most distinguished, who doubt whether the whole universe, or to speak more widely, the whole of being, was only created in Euclid's geometry; they even dare to dream that two parallel lines, which according to Euclid can never meet on earth, may meet somewhere in infinity. I have come to the conclusion that, since I can't understand even that, I can't expect to understand about God. I acknowledge humbly that I have no faculty for settling such questions, I have a Euclidian earthly mind, and how could I solve problems that are not of this world?"[1]

Ivan Karamazov's worries — or perhaps Dostoevsky's — were shared by many people. Yet, although non-Euclidean space was generally considered a strange idea confined to abstract mathematics, there were a few scientists who believed it might be useful in areas of astronomy and cosmology.

3.1.2. ZÖLLNER'S UNIVERSE

Until the 1870s, the hypothesis of non-Euclidean space resulted in a few qualitative predictions, but these were beyond real tests and hence of little scientific interest. Moreover, the hypothesis was apparently unable to throw light on problems of stellar astronomy, which added to its lack of scientific appreciation. In a remarkable work published in 1872, the German physicist and astronomer Karl Friedrich Zöllner demonstrated for the first time

---

[1] https://www.ccel.org/ccel/d/dostoevsky/brothers/cache/brothers.pdf (p. 203).

that the idea of curved space could be used to explain one of the old and much-discussed problems of astronomy known as Olbers' paradox.

Zöllner is today a somewhat obscure figure, but in his own time he was recognized as a brilliant if also controversial pioneer of astrophysics. As a professor in Leipzig, he worked on a variety of subjects including psychophysics, photometry, the physical constitution of the Sun, stellar evolution, and the theory of comets. He also contributed to fundamental physics and to electrodynamics in particular. For example, he suggested an atomic theory with neutral atoms consisting of an equal number of elementary positive and negative charges. Zöllner argued that in this way he could explain Newton's law of gravitation in terms of a net electrical attraction between bodies. Moreover, in this context he established a remarkable numerical relation between electricity and gravitation, two phenomena that apparently had nothing in common. With e and m denoting the charge and mass of the hypothetical electrical particles, and G the constant of gravitation, he derived that

$$\frac{e^2}{Gm^2} \cong 3\times10^{40}.$$

Zöllner did not know about electrons, but in 1904, after the electron had turned into a real and measurable particle, an American physicist by the name Bergen Davis determined independently the value 8×10^{41} for Zöllner's ratio. This very large number — the dimensionless ratio between the electrical and gravitational force — became of interest to theoretical physicists only many years later. The ratio is one of several "large numbers" that have fascinated physicists for almost a century (see Section 1.4.3). There is still no good explanation of why the electromagnetic force is of the order 10^{40} times larger than the extremely weak gravitational force.

Of more relevance to the present issue is a book Zöllner published in 1872, *Über die Natur der Cometen* (On the Nature of Comets), in which he included a chapter wherein he offered an original solution to the old problem of the darkness of the night sky. This problem became eventually known under the eponymous label "Olbers' paradox," a term which first appeared in the scientific literature around 1950. The reference is to the German physician and astronomer Heinrich Wilhelm Olbers, who analyzed the problem in a paper in 1826, but the problem or paradox was known much earlier. It was first noted by Johannes Kepler in a dissertation of 1610. To put it briefly, if we assume an infinite or just hugely large universe uniformly filled with stars in empty space, it can be shown that the starlight received on Earth will make the sky at night brighter than on a sunny day. Of course, this is contrary to experience and therefore needs to be explained or explained away. And yet Olbers, and with him many nineteenth-century astronomers, did not consider it paradoxical at all. Because, as Olbers argued, on the assumption that interstellar space is filled with a rarefied light-absorbing medium, the problem would disappear.

Other astronomers found this explanation to be unsatisfactory and ad hoc, and they consequently looked for alternatives to it. One of the alternatives was to postulate a limited number of stars, but then other problems turned up, in particular that a stellar universe of this kind would be unstable, collapsing gravitationally toward its center of gravity. Yet another way to get rid of Olbers' paradox was to assume a universe of finite age, a solution that Zöllner briefly considered. One could imagine, he wrote, "an act of creation in which had begun, at a time in the finite past, a certain finite initial state of the world" (Zöllner, 1872, p. 306). However, he chose to disregard the possibility of a finite past, which he found unacceptable for philosophical reasons. For him and also for the large majority of his colleagues in astronomy, the idea of a finite-age universe was foreign to the scientific mind. Speculations about cosmic creation were the business of theologians and philosophers, whereas scientists could safely disregard them (see further in Chapter 4.1).

Zöllner argued that the basic problem lay with the generally accepted assumption of infinite space, an assumption which was unnecessary if space were closed rather than flat. As he pointed out, in a Euclidean space filled uniformly with stars, the universe must contain an infinite number of stars, and then Olbers' paradox can only be resolved by means of special hypotheses. But in a universe governed by Riemann's geometry, there will only be a finite number of stars. Zöllner was convinced that the significance of Riemann's discovery of a positively curved space extended far beyond mathematics. It provided nothing less than the key needed to unravel the secrets of the universe and dissolve all problems related to a materially infinite universe (Section 4.2.1). Olbers' paradox was one of the problems, but there were many others, some of which were astronomical and others which related to microphysics. Zöllner (1872, p. 308) wrote: "It seems to me that any contradictions will disappear if we ascribe to the constant curvature of space not the value zero but a positive value, however small. ... The assumption of a positive value of the spatial curvature measure involves us in no way in contradictions with the phenomena of the experienced world if only this value is taken to be sufficiently small."

By means of Riemann's geometry, Zöllner thus explained, at least to his own satisfaction, Olbers' paradox without taking recourse to a limitation of either cosmic time or space. A full explanation of Olbers' paradox is complex and only appeared much later, after the recognition of the expanding universe, but the inadequacy of Zöllner's attempt is beyond the point. What matters is that Zöllner's argument ostensibly provided empirical support for cosmic space being closed and hence finite. This might have been historically important, but unfortunately his argument of 1872 in favor of a Riemannian universe made almost no impact at all on contemporary astronomers and physicists. It was discussed by a few German philosophers but ignored by the astronomers. From their perspective, the hypothesis of a closed universe was unattractive, in part for philosophical reasons and in part because there was neither any need nor observational support for it. The cosmological hypothesis was effectively forgotten and had to be reinvented three decades later.

One of the reasons for the fate of Zöllner's essay was undoubtedly that its author did not follow up on his considerations of a closed-space universe. In none of his later publications did he return to the subject or even refer to his essay. Another reason for the neglect was that Zöllner got a very poor reputation after he, a few years later, crossed the line between science and what most of his colleagues considered to be pseudoscience.

3.1.3. A FOURTH SPACE DIMENSION?

Whereas Zöllner in 1872 endorsed a three-dimensional closed space, a few years later he went on to propose a space of four dimensions. We all know that real space has just three dimensions, neither more nor less, which we take to be self-evident. But why three? Kant was one of the first to seriously consider the question, which he did in a treatise of 1747. According to 23-year-old Kant, the three dimensions followed from Newton's law of gravity, where the force decreases as precisely the square of the distance. With the recognition of non-Euclidean geometries, it was understood that many-dimensional spaces were perfectly possible as mathematical constructs, which allowed for an arbitrary number of dimensions. But this only highlighted the question of why the dimensionality of real or physical space is three. Zöllner thought that four-dimensional space was more comprehensive and, in a sense, more real than the ordinary space of three dimensions. However, his fourth dimension was highly unorthodox as it was as much a spiritual as a physical dimension.

Like several other scientists in the last part of the nineteenth century, Zöllner was deeply interested in spiritualism and the extraordinary, paranormal faculties of particularly gifted individuals. As a result of experiments conducted with an American medium, the famous but also notorious Henry Slade, by the late 1870s he became convinced of the reality behind spiritualist manifestations. In a book in 1879, he reported in great detail his experimental findings, from which he concluded that the spiritual phenomena displayed by Slade and other psychic media must take place in a fourth space dimension unknown to conventional science. It did not occur to him that Slade might be just a skilled conjurer, a cheat who produced what Zöllner wanted to see.

Zöllner's higher aim was to extend physics into what he called a "transcendental physics" comprising material as well as psychical or paranormal phenomena. Contrary to most spiritualists, he emphasized that the spiritual world was not separate from the one of experience. It was subject to scientific investigation and would, for example, satisfy the law of energy conservation. Although the fourth dimension was itself independent of the existence of matter, it governed certain phenomena in the material world that could not be otherwise explained. Thus, Zöllner suggested that without the fourth dimension, the symmetry between three-dimensional bodies could not be understood. He referred to the symmetry between left-handed and right-handed gloves and to the corresponding symmetry between organic molecules known from the new field of stereochemistry.

Not a man of modesty, Zöllner announced his transcendental physics to be a new revolution in worldview comparable to the Copernican revolution. The large majority of contemporary physicists, astronomers, and psychologists found it to be nonsense, closer to pseudoscience than science, and they chose to ignore the bizarre four-dimensional world of the Leipzig professor. Nonetheless, the phantom of the fourth dimension continued to be discussed either outside science or as a potential contribution to scientific insight. According to Simon Newcomb, an eminent American astronomer, the idea was unlikely but not absurd. He speculated that the ether might provide a bridge between our world and the fourth dimension, "For there is no proof that the molecule may not vibrate in a fourth dimension … [and] perhaps the phenomena of radiation and electricity may yet be explained by vibration in a fourth dimension" (Newcomb, 1898, p. 2). Neither Newcomb nor other *fin de siècle* scientists seriously believed in the fourth dimension.

It was sometimes contended that if our three-dimensional space is curved — and this was after all a possibility — it must be contained in a flat space of higher dimension, in the same way that a two-dimensional space is embedded in our three-dimensional space. However, as the mathematicians pointed out (and as Zöllner knew), this is a misunderstanding. A curved space does not need to be curved "in" another space, and generally there is no connection between four-dimensional and curved space. The speculation of a world with other dimensions than three was popularized by the British author Edwin Abbott in his classic novel of 1884, titled *Flatland* and subtitled *A Romance of Many Dimensions*. The novel described the life and experiences of beings living in Flatland, which was a two-dimensional world and not one of four dimensions. Yet, Abbott also referred to the higher view of a four-dimensional world. As he pointed out, as the beings of Flatland are unaware of the third dimension of Spaceland, so we are unaware of the hidden fourth dimension of our experienced world:

> "It is true that we have really in Flatland a Third unrecognized Dimension called 'height', just as it is also true that you have really in Spaceland a Fourth unrecognized Dimension ... which I will call 'extra-height'. But we can no more take cognizance of our 'height' than you can of your 'extra-height'. ... Even I cannot comprehend it, nor realize it by the sense of sight or by any process of reason; I can but apprehend it by faith" (Abbott, 1998, p. 2).

About two decades later, the fourth dimension reappeared as a theme in modernist art, and that even before Einstein famously added time as an extra dimension to the three space dimensions. Cubist painters of the early twentieth century, such as Pablo Picasso and Georges Braque, sought to incorporate the four-dimensional perspective in their art. With some delay, Einstein's theory of relativity made considerable impact on the artistic and literary world, where the fourth time dimension was frequently misunderstood as an extension of space to four dimensions.

The space-time of relativity theory is four-dimensional, but space still has only three dimensions and time one. In classical physics, the square of the distance between two neighbor points in space, (x_1, x_2, x_3) and $(x_1 + dx_1, x_2 + dx_2 + x_3 + dx_3)$, is given by the Euclidean formula $ds^2 = dx_1^2 + dx_2^2 + dx_3^2$. The corresponding formula in the theory of special relativity is the Minkowski metric named after the mathematician and theoretical physicist Hermann Minkowski. It can be written in a conspicuously similar form

$$ds^2 = dx_1^2 + dx_2^2 + dx_3^2 + dx_4^2$$

suggesting a simple extension from three to four dimensions. However, the last term is here an abbreviation of $-c^2dt^2$ or $dx_4 = icdt$, where i denotes the imaginary unit $\sqrt{-1}$. Thus, the physical meaning of the temporal term is different from that of the three spatial terms.

Does the mathematical possibility of higher space dimensions correspond to anything in the real world investigated by physicists? Zöllner thought so, but apart from being plain wrong he was also alone in his belief. As the philosopher Bertrand Russell, who in his youth specialized in the foundation of mathematics, remarked in a treatise of 1897, "With the exception of Zöllner, I know of no one who has regarded the fourth dimension as required to explain phenomena" (Russell, 1897, p. 53).

Quite independent of Zöllner's fantasy of a transcendental physics, the fourth space dimension returned in about 1920, when a few mathematicians and physicists tried to unify Maxwell's theory of electromagnetism and Einstein's theory of gravitation. The German mathematician Theodor Kaluza presented in 1919 a five-dimensional extension of Einstein's four-dimensional space-time, and a couple of years later the Swedish physicist Oskar Klein independently proposed a fourth space dimension in an even more ambitious unified theory that included quantum phenomena. The extra dimension appearing in what was later called Kaluza–Klein theory was not directly observable. As Klein argued, it was "curled up" or "compactified" in a loop with a radius as incredibly tiny as 10^{-32} m or thereabout (in comparison, the radius of a proton is about 10^{-15} m). What is sometimes called the Klein length l_K is of roughly the same order as the Planck length l_P and can be expressed in terms of this quantity as $l_K = 4\pi\sqrt{2/\alpha}\, l_P$, where α is the fine-structure constant (see Section 1.4.1).

Although the original ideas of Kaluza and Klein turned out to be unfruitful, they provided the basis for further research on unified and many-dimensional physical theories. The best known of these modern theories is the theory of superstrings, which operates with 9 or 10 space dimensions and thus 6 or 7 more than in conventional physics. The number of extra dimensions are not required for physical reasons, but because they are needed for string theory to be mathematically consistent. Do these other dimensions exist elsewhere than in the minds of string physicists? The short and honest answer is that we do not know. There is so far no convincing evidence for either the fourth space dimension or any higher dimensions, and yet it is possible that such evidence will turn up in future experiments at very high energy. According to some modern theories, extra dimensions may also be large and show up in studies of gravitational waves. But again, no evidence has been found and so astronomers are confident that the universe at large has only the dimensions we are used to.

3.1.4. FIN DE SIÈCLE ASTRONOMY

But back to history, and back to non-Euclidean geometry. Very few astronomers found it worthwhile to discuss the possibility of curved cosmic space, and even fewer considered it more than an exotic hypothesis. Newcomb was thoroughly acquainted with non-Euclidean geometry, which he examined in several books and articles. In the widely read *Popular Astronomy*, a book first published in 1878, he discussed some of the astronomical consequences of a finite and positively curved universe. One of them concerned the question of what happened with the enormous output of solar heat. If space were flat or hyperbolic, the radiant heat would presumably be lost forever, implying that the Sun had a finite and relatively short lifetime (see also Section 3.2.1). However, if space were closed, the heat might eventually return to the Sun after a very long cosmic journey. Newcomb found this an intriguing possibility, but no more than that. After having mentioned the possibility of Riemann's closed space, he concluded that the hypothesis was "too purely speculative to admit of discussion" (Newcomb, 1878, p. 505). In agreement with the consensus view, Newcomb suggested that solar heat was irretrievably lost in infinite space.

The idea of curved astronomical space received unqualified support only from a single scientist, and this scientist is today much better known as a philosopher and founder of semiotics, the study of signs. The polymath Charles Sanders Peirce, a friend and correspondent of Newcomb, was trained in astronomy and spent most of his professional life as a practicing scientist. As pointed out in Section 2.2.3, based on spectroscopic studies he suggested the existence of a chemical element in the upper atmosphere with atomic weight less than that of hydrogen. For a decade, Peirce enthusiastically defended the hypothesis of curved space, which he endeavored to prove by means of astronomical data. Peirce was convinced of the significance of his research program, such as he told Newcomb in a letter in 1891:

> "The discovery that space has a curvature would be more than a striking one; it would be epoch-making. It would … lead to a conception of the universe in which mechanical law should not be the head and centre of the whole. … In my mind, this is part of

> a general theory of the universe, of which I have traced many consequences, — some true and others undiscovered, — and of which many more can be deduced; and with one striking success, I trust there would be little difficulty in getting other deductions tested. It is certain that the theory if true is of great moment" (Eisele, 1957).

While Zöllner and Newcomb focused on the closed space, Peirce thought that the hyperbolic space of Lobachevsky was in better agreement with astronomical data. In manuscripts and notes, most of them unpublished, he carefully discussed methods of investigating the curvature of space. These included not only parallax measurements but also the proper motions of stars, stellar evolution, and counts of stars of different magnitudes. While at first optimistic that he could provide solid evidence for a negatively curved space, after a few years he was forced to admit that his optimism was unfounded. In any case, his heroic defense of curved space made almost no impact since his arguments were not reported in journals read by the majority of astronomers. When Peirce died in 1914, observational proof of curved space was still lacking.

The general opinion at the turn of the century was summarized by Bertrand Russell in a work on the foundation of geometry. "Though a small space-constant is regarded as empirically possible," Russell wrote, "it is not usually regarded as probable; and the finite space-constants with which Metageometry is equally conversant, are not usually thought even possible, as explanations of empirical fact" (Russell, 1897, p. 53). While many astronomers agreed that the geometry of physical space could in principle (and only in principle) be determined by observations, according to the influential French mathematician and physicist Henri Poincaré this was not the case. As he saw it, the geometry of space was not a measurable objective property but just a matter of convention. One geometry might be more convenient than another, but it could not be more true. From Poincaré's conventionalist point of view, the question of the real structure of space was a pseudoquestion. He found the Euclidean geometry to be simple and convenient, and therefore saw no reason to consider other candidates for the structure of space.

The period's most elaborate attempt to provide a link between astronomy and non-Euclidean geometry was due to the 26-year-old German astronomer Karl Schwarzschild, who would later be celebrated for his pioneer analysis of what today is called black holes. In a lecture in 1900, Schwarzschild systematically discussed how to determine the geometry of space from astronomical observations. He thus addressed the same problem as first considered by Lobachevsky and later by Peirce, but Schwarzschild's analysis was more complete and relied on better data. Nonetheless, he was unable to go much further than his predecessors.

In the case of a hyperbolic space, Schwarzschild found from parallax measurements that the radius of curvature must exceed 4×10^9 AU or 64,000 light years. For the closed Riemannian case, he estimated a lower bound of $R = 1.6 \times 10^8$ AU = 2500 light years. Schwarzschild saw no way to go further than these rather indefinite bounds or to decide whether space really has a negative or positive curvature. In spite of this agnostic conclusion, on philosophical grounds he preferred the latter possibility corresponding to a finite universe. Because, "if this were the case, then a time will come when space will have been investigated like the surface of the Earth, where macroscopic investigations are complete and only the microscopic ones need continue" (Schwarzschild, 1900). Schwarzschild was of course fully aware that his emotional preference of a closed universe did not count as a scientific argument.

During the early years of the twentieth century, the astronomers' favored view of the stellar universe was essentially limited to the Milky Way system. Observations and complex calculations indicated that the system was of an ellipsoidal form, with the density of stars diminishing with increasing distance from the center. The dimensions were of the orders 50,000 light years in the galactic plane and 5000 light years towards the galactic poles. The huge conglomerate of stars in this system was often pictured as floating in an infinite Euclidean space, either empty or filled with some kind of ether, but the space beyond the stellar system was left for speculation and considered irrelevant from an astronomical point of view.

Inspired by Schwarzschild, in 1908 the German astronomer Paul Harzer, a professor at the University of Kiel, suggested a modified picture based on the assumption of a closed space. In this way, he did away with the problematic infinite space of the traditional models, arriving at a smaller stellar universe placed in a spherical space approximately 20 times the size of the Milky Way. Taking absorption into consideration, he estimated that the time it would take a ray of light to circumnavigate the entire universe was of the order 9000 years. Although Harzer took his model of the closed stellar universe seriously, he realized that it was just a hypothesis unsupported by observational evidence.

The Schwarzschild–Harzer suggestion of a closed space filled with stars left little impact on mainstream astronomy. By 1910 — nearly a century after Gauss had aired the idea of non-Euclidean space — astronomers still saw no reason to abandon the traditional flat space that had served them so well in the past. In a nutshell, they had no need for the extravagant hypothesis of curved space. A decade later, the situation changed, not because of better astronomical data but because of Einstein's general theory of relativity and the intimate connection between gravity and space curvature posited by this theory. According to all earlier theories, the structure of space was an intrinsic property and not one that could be affected locally by the gravitational field of massive bodies. Ever since Newton, gravitation was recognized to be of crucial importance in astronomy and cosmology, but until Einstein came along no one thought of relating the force of gravity to the structure of space.

3.1.5. EINSTEIN'S CURVED SPACE

The positive curvature of space as hypothesized by Riemann in 1854 was discovered as a real phenomenon 65 years later. To be precise, on May 29, 1919.

Until the early 1910s, Einstein did not pay attention to non-Euclidean space, a branch of geometry he was only vaguely aware of and did not consider important to his theory of relativity. He was at the time contemplating how to extend the original theory to cover gravitation as well, and in this context he formulated what he called a general principle of equivalence. According to this principle, no mechanical experiment can distinguish between a homogeneous gravitational field and a uniformly accelerated frame in which there is no gravitational field. A constant acceleration acts as gravity, albeit in the opposite direction. In 1911, Einstein deduced from the equivalence principle the remarkable result that the path of a ray of light would not necessarily be straight, as in Euclidean space. No, in the vicinity of a massive body of mass M the light would but be deflected towards the body. For light grazing the body in the distance D, Einstein found for the angle of deflection

$$\varphi = \frac{2GM}{Dc^2}.$$

In the case of the Sun, with D equal to its radius, the deflection would be $\varphi = 0.87''$, a very small but possibly measurable quantity. "It would be a most desirable thing if astronomers would take up the question here raised," he commented (Einstein *et al.*, 1952, p. 108).

Einstein's space or rather space-time of 1911 was still flat, with no trace of Riemann's geometry. It was only the following year, due to a collaboration with his friend Marcel Grossmann, a professor of mathematics, that Einstein realized the crucial importance of non-Euclidean space-time. According to the new theory, space-time was not a fixed background on which physical events unfolded, it was a dynamical or plastic quantity varying with the gravitational field. The equations that Einstein and Grossmann announced in 1913 led to the same result for the gravitational bending of light as in 1911, but it was now interpreted as due to the geometry of space being locally distorted by the presence of a massive body. Already at that time Einstein had the insight that it might be possible to detect the effect during a solar eclipse, and astronomers began looking for it.

The Einstein–Grossmann theory turned out to be untenable for both conceptual and empirical reasons. It took another two years of hard work until Einstein, in November 1915, could present his general theory of relativity in its final and still accepted form. The new theory retained the result of light bending obtained in the earlier theory, but now the predicted angular deflection was twice as large, $4GM/Dc^2$ instead of $2GM/Dc^2$, which for the Sun resulted in $\varphi = 1.74''$. This was a novel and most radical prediction, which if verified would prove curved space to be real and not a mathematical fantasy.

It was realized early on that only during a total solar eclipse would it be possible to detect the bending of starlight as it passed the rim of the Sun. The first attempts failed or were inconclusive, and the outbreak of World War I prevented solar eclipse expeditions for more than four years. It was only with the famous British expedition of May 1919, headed by Frank Dyson and Arthur Eddington, that good data were obtained for background stars which would be invisible — hidden behind the Sun — if space were flat. While Dyson stayed in London, Eddington went with a team to Principe Island off the West African coast, and another team took photographs at Sobral in northern Brazil. After an elaborate analysis of the photographic plates, the British astronomers reported a light deflection in agreement with Einstein's prediction. The Sobral measurements gave 1.98 ± 0.16 and the less convincing Principe Island measurements 1.61 ± 0.40, both values in seconds of arc.

Although the agreement between theory and observation was contested in some quarters of science, generally the results of the eclipse expedition were hailed as a triumph and even proof of Einstein's theory. Not only had the Swiss-German physicist proved Newton wrong, he had also disproved the age-old belief of a flat Euclidean space. As Newton had erred, so had Euclid and Kant. Einstein received the good news of Eddington's verification in September 1919 and immediately wrote to his mother, Pauline Einstein, that "the English expeditions have definitely proved the deflection of light by the Sun" (French, 1979, p. 102). As to Eddington, he described the Principe expedition in the form of a poem, with the last verse reading:

> "Oh leave the Wise our measures to collate
> One thing at least is certain, light has weight
> One thing at least is certain, and the rest debate —
> Light-rays, when near the Sun, do not go straight" (Douglas, 1956, p. 44).

The claim that light has weight may seem curious, for the photons that comprise light are absolutely massless. Not only does this follow from theory, present experiments gives an upper limit of 10^{-18} eV or less than the mass of an electron by the huge factor of 10^{24}.

Nonetheless, light traveling in a gravitational field can metaphorically be ascribed a weight, if not, strictly speaking, a mass. Eddington was fond of transcribing curved space-time to weight of light.

The demonstration of 1919 that space near a heavy celestial body is positively curved was extremely important, but it was irrelevant with regard to the structure of space on a cosmological scale, the problem that had been considered by earlier astronomers from Lobachevsky over Zöllner to Schwarzschild. Stimulated by suggestions of Schwarzschild and the Dutch astronomer Willem de Sitter, in the autumn of 1916 Einstein began thinking of how to apply general relativity to the universe as a whole. The next year, at a meeting in Berlin of 8 February, he presented a path-breaking model that initiated a new and immensely fruitful chapter in the history of cosmology. In order to maintain a static equilibrium universe in accordance with observations, Einstein felt forced to modify his field equations of 1915, which he did by introducing a "cosmological constant." The new constant represented a cosmic repulsion proportional to distance, but its value was unknown except that it must be very small and equal to the space curvature. Much later, the cosmological constant usually designated with the symbol Λ was interpreted as a measure of the vacuum energy density or dark energy responsible for the accelerated expansion of space. More about the cosmological constant and Einstein's models of the universe follow in Chapters 4.2 and 4.4.

Einstein's equations of 1917, known as the cosmological field equations, have a canonical status today. Einstein mistakenly believed that there was only one physically realistic model consistent with the field equations, and that this model, apart from being static, was finite, filled with dilute matter, and spatially closed in the sense of Riemann's geometry. "The curvature of space is variable in time and space, according to the distribution of matter," he wrote, "but we may roughly approximate to it by means of a spherical space" (Einstein *et al.*, 1952, p. 188). For the curvature, or the inverse square of the curvature radius, Einstein derived the expression

$$K = \frac{1}{R^2} = \rho \frac{4\pi G}{c^2},$$

where ρ denotes the average density of matter. The total mass of Einstein's universe was given by

$$M = 2\pi^2 R^3 \rho.$$

With the much too high estimate $\rho = 10^{-22}\,\mathrm{g/cm^3}$, the radius came out as no more than 10 million light years, values that Einstein suggested in private but not in public. In 1926, Edwin Hubble used Einstein's model and the density $\rho = 1.5 \times 10^{-31}\,\mathrm{g/cm^3}$ to derive a radius of the universe equal to 10^{11} light years. Although Einstein's model was wrong, for other reasons because it presupposed a static universe, it built on the right equations and is the legitimate grandfather of all later models. For a decade or so, the closed Einstein model was widely regarded as the most credible picture of the universe. When it was realized in the early 1930s that the universe is expanding, the static model was discarded but the cosmological equations were left intact (see Section 3.4.1).

In an address to the Prussian Academy of Science in early 1921, Einstein reflected on the relationship between mathematics and the physical sciences. He famously stated that "as far as the propositions of mathematics refer to reality, they are not certain; and as far as they are certain, they do not refer to reality" (Einstein, 1982, p. 233). Moreover, he distinguished between what he called "practical geometry" and "purely axiomatic geometry," arguing with Gauss that while the first version was a natural science, the second was not. "The question whether the universe is spatially finite or not seems to me an entirely meaningful question in the sense of practical geometry," he said. "I do not even consider it impossible that the question will be answered before long by astronomy."

However, like Peirce some twenty years earlier, Einstein's cautious prophecy was no more than wishful thinking. Reliable empirical methods to determine the curvature constant k were only developed many years after the discovery of the expanding universe, and for a long period of time measurements were too crude to distinguish between flat ($k = 0$), closed ($k = +1$), and hyperbolic ($k = -1$) space geometries. Present precision cosmology built on measurements of the average matter density, the Hubble constant, and other cosmological parameters fares much better. These measurements, coupled with theoretical predictions of the standard cosmological model, suggest that space is flat or very nearly so. Many cosmologists believe that we live in an infinite universe (Section 4.2.4). And yet, we do not know, and still today it is worth recalling what Lobachevsky pointed out back in 1829, namely that it can never be proved observationally that cosmic space has the precise curvature constant $k = 0$.

3.2

Early Theories of Stellar Energy

3.2

Early Theories of Stellar Energy

Every second, the Sun pours out an enormous amount of heat and light, and has done so for hundreds of million years. A tiny fraction of the Sun's energy hits the surface of the Earth, making life possible. What is the cause of sunlight and of starlight generally? In his classic work, *Treatise on Astronomy* dating from 1833, the distinguished British astronomer and natural philosopher John Herschel dealt with the question of the source of the Sun's heat and light. "Every discovery in chymical science here leaves us completely at a loss, or rather, seem to remove further the prospect of a probable explanation," he despairingly wrote (Herschel, 1833, p. 202). Whatever the Sun was, it was not a chemical machine. Burning of coal or other combustion processes simply would not do, and neither would ad hoc hypotheses of mechanical friction or electrical discharges. Herschel concluded that "we know nothing, and we may conjecture every thing."

A decade later, with the new concept of energy as a conserved quantity, it looked as if the Sun's outpouring of heat was mechanical in origin. For a brief while, gravitational capture of meteors was considered a possible mechanism. During a period of more than forty years, from about 1860 to 1905, the standard model of solar energy production was slow gravitational contraction, a theory first proposed by Helmholtz and subsequently developed by William Thomson. The Helmholtz–Thomson contraction theory was highly regarded by astronomers and physicists, notwithstanding that it only provided an age of the Sun of some 20 million years.

By the turn of the century, confidence in the contraction theory decreased, which, after a brief interlude where radioactivity was considered as a possible alternative, resulted in the first serious proposals that solar energy had its origin in subatomic processes. At the time, physicists knew, or thought that they knew, that the atomic nucleus consisted of protons and electrons, and many believed that somehow the two elementary particles must be responsible for the Sun's heat. Two possibilities were discussed during the 1920s, either proton–electron annihilation or the exothermic fusion of hydrogen atoms into helium atoms. Still, by the late 1920s, solar energy resisted explanation in terms of accepted physics, which caused Niels Bohr to suggest that the law of energy conservation was violated in the interior of stars. Bohr's proposal was radical, indeed much too radical to win support. The problem of the Sun's heat was difficult but not believed to be unsolvable, such as indicated by the first attempts to treat thermonuclear stellar reactions by means of quantum mechanics. These theories dating from 1929–1931 were promising, and yet they did not provide a final answer. One of the reasons, of course unknown at the time, was that the atomic nucleus was still considered to consist of protons and electrons.

3.2.1. THE ATTRACTION OF GRAVITATION

In the early part of the nineteenth century, astronomers knew next to nothing about the physics and chemistry of the Sun. About the only physical property known with some confidence was the Sun's average density, which Newton had derived by a remarkable chain of reasoning in his *Principia* to be approximately four times smaller than the density of the Earth. With Henry Cavendish's experimental determination of the mass of the Earth, dating from 1798, the Sun's average density came out as $1.37\,g/cm^3$, close to the modern value of $1.41\,g/cm^3$.

The Sun is a hot body, but how hot is it? How much heat does it emit per unit of time? The latter question refers to what is known as the solar constant, which is a measure of the solar energy received at the surface of the Earth outside its atmosphere. Today, the quantity is known with great accuracy to be 1.361 kW/m^2. The credit for the first reliable measurement of the solar constant goes to the French physicist Claude Pouillet, director of the Conservatoire des Arts et Métier in Paris, who invented a special kind of calorimeter designed to measure the heating effect of sunlight. In a paper published in 1838, Pouillet coined the name "solar constant" and reported its value to be 1.76 cal/cm^2/min (calories per square centimeter per minute) or, in modern units, 1.23 kW/m^2. The recognition of the Sun's enormous output of heat naturally provoked the question of how the heat was generated, a question that could only be addressed on a scientific basis after the law of energy conservation was established a few years later (Section 1.1.1). While previously it had made sense to conceive the Sun as an eternal cosmic object, it now became imperative to discuss the source of the Sun's heat and also the time at which it would cease to shine, if ever.

The first to address the question was the chief discoverer of energy conservation, Robert Mayer, who was aware of Pouillet's determination of the solar constant and realized that chemical processes were hopelessly inadequate to provide the necessary heat energy. In a privately printed paper from 1848 titled "Beiträge zur Dynamik des Himmels" (Contributions to Celestial Dynamics), Mayer suggested that solar activity was gravitational in origin, namely, due to a constant falling of meteors or asteroids into the Sun. He argued that a meteor striking the Sun at a speed of approximately 500 km/s would develop a heat energy more than 7000 times as much as the heat generated by the combustion of an equal mass of coal. Mayer thought that each minute the Sun might swallow a mass of the order of 10^{14} kg and that the radiated energy would in this way be fully compensated, leaving the Sun in a steady state. Contrary to other contemporary scientists, he believed that the Sun's present activity would never cease.

Mayer's paper, the last of his scientific publications, was little known until it was translated into English in 1863. Thus, William Thomson was unaware of it when he, in 1854, presented his own version of the meteoric-gravitational hypothesis, which assumed an influx of no less than 100 Earth masses every 4750 year, or 2×10^{17} kg per minute. For nearly a decade, Thomson defended versions of the meteoric hypothesis, but it never won general recognition. The hypothesis presupposed an extraordinary number of heavy meteors or asteroids currently falling into the Sun, an assumption for which there was evidence at all. During around 1862, Thomson concluded that a much better alternative was to be found in another gravitational mechanism, a slow shrinking of the Sun, which was first proposed by Hermann Helmholtz. This theory, known as the Helmholtz–Thomson or Helmholtz–Kelvin theory (and today as the KH mechanism), came to be celebrated as the authoritative and probably correct theory of solar energy production during the late Victorian period. Its essence was that the Sun contracted slightly and the gravitational energy released during the process manifested itself in the liberation of light and heat.

In an important lecture of February 1854 given in Königsberg, Germany, Helmholtz argued that even an inappreciable contraction of the Sun would result in an energy corresponding to the solar heat radiation over a period of more than a thousand years. For the potential gravitational energy of the Sun, for simplicity supposed to be of uniform density, he calculated what is equivalent to the modern expression

$$V = -\frac{3}{5}\frac{GM^2}{R},$$

where M and R refer to the Sun's mass and radius, respectively. From this and the arbitrary assumption that the specific heat capacity of the Sun is the same of that of water, Helmholtz concluded that if R were diminished by the ratio $1:10^4$ of its present value, it would generate an increase in the Sun's temperature of about 2900°C. Taking into account Pouillet's solar constant, the slight contraction would correspond to the Sun's energy output during a period of about 2300 years. Helmholtz did not estimate the age of the Sun in his Königsberg lecture, but he did so in a later address in 1871 in which he stated the age to be approximately 22 million years. Moreover, he looked into the future fate of the Sun:

> "We may therefore assume with great possibility that the Sun will still continue in its condensation, even if it only attained the density of the Earth ... this would develop fresh quantities of

> heat, which would be sufficient to maintain for an additional 17 000 000 of years the same intensity of sunshine as that which is now the source of all terrestrial life" (Helmholtz, 1995, p. 270).

All the same, in the end the Sun would become extinct and with it all life (see also Section 1.1.5). Thomson agreed and further developed the contraction theory in a series of works. For example, he took into account the increase of the Sun's density toward the center and also used the latest and presumably best values of the solar constant. In the late 1880s, he adopted the American astronomer Samuel Langley's too large value of $2.14\,\mathrm{kW/m^2}$, from which he concluded that the Sun could not have existed as a strongly radiating body for more than 20 million years and that it would cease to shine appreciably in another 5 or 6 million years.

During the last decades of the nineteenth century, the Helmholtz–Thomson contraction theory was generally recognized as true by the majority of physicists and astronomers. The two names associated with the theory, widely considered the greatest theoretical physicists of their time, added to its authority. Although several scientists proposed alternatives to the gravitational contraction theory and its pessimistic message of a dying Sun, none of them were taken very seriously. And yet, although the Helmholtz–Thomson theory was the accepted standard model of solar energy production, it was widely recognized that it was not entirely satisfactory. For one thing, it rested on a phenomenon — the gradual shrinking of the Sun — which lacked observational confirmation. Thomson calculated that the Sun's diameter would decrease with 70 m per year, but careful studies made by astronomers showed no trace of the predicted diminution. As far as observations could tell, the size of the Sun remained constant. Generally, the Helmholtz–Thomson theory was difficult to test, if testable at all.

More important and much more discussed was the theory's prediction of a lifetime of the Sun, which clashed with the age of the Earth as advocated by geologists and natural historians. The age of the Earth could conceivably but not realistically be shorter than the age of the Sun, and geologists claimed that the Earth was at least 100 million years old and probably as old as 300 million years. By the turn of the century, it was generally recognized that Thomson's value of the order of 20 million years was much too small, and yet the contraction theory lived on for another decade or two if only as a shadow of its former glory. As Eddington memorably phrased it in an address in 1920: "If the contraction theory was proposed to-day as a novel hypothesis I do not think it would stand the smallest chance of acceptance. ... Only the inertia of tradition keeps the contraction theory alive — or, rather, not alive, but an unburied corpse" (Eddington, 1920b). He further stated that, "Lord Kelvin's date of the creation of the sun is treated with no more respect than Archbishop Ussher's," a reference to James Ussher's notorious conclusion from 1650 that God had created the world on 22 October 4004 BC.

By the time of Eddington's address, it was vaguely recognized that the Sun's production of energy must be explained in terms of interacting subatomic particles, an idea that was proposed in a preliminary version just a few years after the discovery of radioactivity in 1896.

3.2.2. THE RADIOACTIVE SUN

Radioactivity attracted wide interest, scientifically as well as publicly, only after 1898 when Marie and Pierre Curie discovered radium, a rare but magic element much more active than uranium. By the early years of the new century, Ernest Rutherford and his collaborator Frederick Soddy had established that radioactive substances decay spontaneously and apparently randomly into atoms of other elements. It was as if nature had fulfilled the alchemists' dream of element transmutation. The strange phenomenon became even stranger when Pierre Curie and his assistant Albert Laborde in 1903 reported their discovery of an enormous heat energy released by radium and its decay products. In their delicate experiment with no more than 0.08 g of pure radium chloride, they measured the energy using an ice-calorimeter, finding that the spontaneous process resulted in about 100 calories per gram per hour, or about 200,000 times more than the total burning of coal.

The researchers in Paris did not associate the radioactive heat with the heat radiation emitted by the Sun, but other scientists did. According to Rutherford, an estimate based on the Curie–Laborde value and Langley's value of the solar constant allowed the conclusion that "the presence of radium in the sun, to the extent of 2.5 parts by weight in a million, would account for its present rate of emission of energy" (Rutherford, 1904, p. 343). The Irish astronomer Robert Ball, professor of astronomy and geometry at Cambridge University, had been convinced of the Helmholtz–Thomson contraction theory while at the same time recognizing its weaknesses. He now eyed in radioactivity a mechanism that might better explain the Sun's prodigious generation of energy. His friend and compatriot, John Joly, a physicist and geologist at Trinity College, Dublin, was deeply involved in the controversy over the age of the Earth and in favor of a time-scale longer than Thomson's. In a letter to Joly in 1903, Ball enthusiastically wrote:

> "Have you seen radium? It certainly gets over the greatest of scientific difficulties, viz. the question of sunheat. The sun's heat cannot have lasted over 20 000 000 years if it is due to contraction. But the geologist would have, say, 200 000 000. Now the discrepancy vanishes if the sun consists in any considerable part of radium, or something that possesses the like properties. It is a most instructive discovery" (Hufbauer, 1981).

For a decade or so, the cosmic role of radioactivity was much discussed by physicists and astronomers in connection with the problem of stellar energy. The new phenomenon also turned up in a cosmological context, namely, in the discussion of whether the universe was eternal or had only existed for a finite number of years (Section 4.1.1). Radioactivity seemed to be an argument for the latter option. Among the reasons for assuming the Sun to be a radioactive machine was the recent discovery of terrestrial helium, the element that was originally identified as a constituent of the Sun's atmosphere. Due to the work of Rutherford in particular, by 1905 it was known that helium in the form of alpha particles ($\alpha = He^{2+}$) was a decay product of radioactive change, which made it tempting to infer that the Sun's content of helium derived from radioactive sources in its interior. Unfortunately, nothing was known about the abundance of solar helium.

For a period of time, the radioactive hypothesis was popular and received qualified support by some of the period's most eminent scientists, among them Rutherford, Soddy, Johannes Stark, and Henri Poincaré. On the other hand, the aging William Thomson would have nothing to do with it and maintained until his death in 1907 that the gravitational contraction theory was able to account for all relevant facts related to both the Sun and the Earth. We have in the famous writer H. G. Wells, the author of *The Invisible Man* and much more, an interesting example of how a non-scientist received the news of a possibly radioactive Sun. In 1902, Wells gave a lecture to the Royal Institution on "The Discovery of the Future" in which he reflected on the consequences of the Helmholtz–Thomson theory, namely, that the Sun would eventually become extinct and all life on Earth would vanish. When the lecture was reprinted eleven years later, he added a footnote in which he pointed out that the discovery of radioactivity had radically changed the prospect.

The hypothesis of a radioactive Sun was, to a large extent, based on wishful thinking rather than facts, and it was never developed into a proper theory based on measurements and with definite predictions. Given the vagueness of the hypothesis, it was not possible to calculate the radioactive energy flux and compare it to the solar constant without arbitrary assumptions. Moreover, it was obviously a problem that spectra of the solar atmosphere failed to reveal lines of radium or other radioactive elements. Although there were a few claims to have detected spectral lines of radium, uranium, and radon (at the time called radium emanation), the claims were disputed and soon realized to be unfounded.

Another weakness concerned the radioactive elements that supposedly powered the Sun. Radium with a half-life of only 1600 years (Ra-226) was insufficient and so was the long-lived but only weakly radioactive uranium. As an alternative, one might fall back on speculations, such that ordinary elements in the intensely hot Sun decomposed to alpha-active elements similar to radium but with a much longer lifetime. Something like this is what Rutherford suggested in his 1904 monograph *Radio-Activity*: "It is not improbable that, at the enormous temperature of the sun, the breaking up of the elements into simpler forms may be taking place at a more rapid rate than on the earth. If the energy resident in the atoms of the elements is thus available, the time during which the sun may continue to emit heat at the present rate may be from 50 to 500 times longer than was computed by Lord Kelvin from dynamical data" (Rutherford, 1904, p. 344). Rutherford repeated the speculation in his updated monograph *Radio-Active Substances and Their Radiations* from 1913, but without elaborating or making it more convincing.

The radioactive solar hypothesis was nothing more than a parenthesis in the history of astrophysics, proposed about 1903 and abandoned some twelve years later. Although a failure, it can be considered a precursor to the new ideas of solar and stellar energy based on subatomic processes that were developed in the early 1920s and to which we shall now proceed.

3.2.3. ANNIHILATION AND FUSION HYPOTHESES

By the late 1910s, the Rutherford–Bohr model of the atom was universally accepted, including its picture of the atomic nucleus as a densely packed conglomerate of protons and electrons. During the following decade, it became increasingly evident that somehow the energy of the stars was due to reactions between subatomic particles, although it was far from clear which reactions were responsible. Three kinds of mechanisms were discussed: (1) proton–electron annihilation; (2) fusion of hydrogen to helium; (3) radioactive decay of hypothetical transuranic elements. By the end of the 1920s, the second possibility seemed the most likely one, but it was just a qualitative hypothesis without solid support in either theory or experiment. The mechanism governing stellar energy generation was as mysterious as ever.

In a paper published in 1917, Eddington briefly suggested that a stellar time-scale much longer than the one allowed by the contraction theory might be provided by a slow annihilation of protons and electrons, that is, $p^+ + e^- \rightarrow E$, where E represents some form of radiation energy. Eddington actually referred to the annihilation of "positive and negative electrons," but there is little doubt that he conceived the positive electron to be a hydrogen ion H^+ or proton (a name which was only coined in 1920). The annihilation hypothesis was not Eddington's invention, for it had been introduced by James Jeans as early as 1901, at the time with the positive electron referring to a hypothetical mirror particle of the negative electron (Section 1.3.1). However, Eddington was the first to consider proton–electron annihilation and apply the idea to the problem of stellar energy production. Jeans returned to annihilation in 1924 and over the next several years promoted it in a series of papers and public addresses. For example, in a lecture in 1928 he stated confidently that "the falling together of electrons and protons forms the obvious mechanism for the transformation of matter into energy, and it now seems practically certain that this is the actual source of the radiation of the stars" (Jeans, 1928b).

Eddington was not so sure as Jeans, and in an address of 1920 to the British Association for the Advancement of Science he suggested as another possibility that the energy of a star was released when helium nuclei were produced by a fusion of four hydrogen atoms. The process he imagined was

$$4\,{}^{1}_{1}\mathrm{H} + 2\,{}^{0}_{-1}e \rightarrow {}^{4}_{2}\mathrm{He}.$$

Eddington considered the fusion scenario attractive because it provided a unitary explanation of energy production and the synthesis of elements. He was one of the first to propose that stellar energy goes hand in hand with the formation of heavier elements. Moreover, the fusion process seemed promising because it was theoretically justified in the known atomic masses of the proton and the helium nucleus. According to Francis Aston's recent experiments with the new mass spectrograph, the mass of a helium nucleus was about 0.7% less than that of four protons, which implied (because of Einstein's $E = Mc^2$) that the hypothetical reaction 4H → He would be strongly exothermic. In his address of 1920, Eddington stated optimistically that the stellar energy problem would be solved if only 5% of a star's mass consisted of hydrogen atoms, and these were transformed into helium and from there possibly to more complex elements.

The idea of helium fusion from hydrogen was not Eddington's invention, although he may have arrived at it independently. It had first been suggested by the American chemist William Harkins, who in 1917 suggested the 4H → He process as the first step in the formation of complex atoms. While Harkins did not relate his suggestion to stellar processes, this is what Jean Perrin, a French physicist and future Nobel laureate of 1926, did in papers of 1920–1921. According to Perrin, solar energy might have its source in helium synthesis. Although his ideas were roughly similar to Eddington's, they were not well known and made no impact on astronomers.

Without clearly preferring one of the two alternatives, annihilation and helium synthesis, Eddington compared them in a chapter in his influential monograph of 1926, *The Internal Constitution of the Stars*. As he fully realized, both alternatives were problematic. Among the physical difficulties facing the fusion hypothesis was that it required the extremely improbable collision of six particles, which would only be possible if they moved at enormous velocities corresponding to a temperature in the interior of stars far exceeding the 40 million degrees that astronomers at the time estimated for the Sun. Eddington further pointed out that red giant stars presented a problem, which he illustrated by comparing the Sun with the red giant Capella. It was known that the Sun was considerably hotter than Capella and about 600 times as dense, and yet the energy output of Capella was 58 erg/g/s, much more than the 1.9 erg/g/s emitted by the Sun. This was inexplicable if the stars were powered by the H-to-He fusion process.

Unable to come up with an explanation of the red giant paradox (as he called it), an apparently undisturbed Eddington countered this and other difficulties by means of rhetoric: "But the helium which we handle must have been put together at some time and some place. We do not argue with the critic who urges that the stars are not hot enough for this process; we tell him to go and find a *hotter place*" (Eddington, 1926, p. 301). With the benefit of hindsight, the hotter place imagined by Eddington can be identified with the big bang, during which almost all helium in the universe was produced, but this was not what Eddington had in mind. He never accepted that the universe originated in a big bang.

The annihilation hypothesis was even more problematic as it was purely speculative and unsupported by experiments. In the version of Jeans, who assumed a single photon to be emitted by $p^+ + e^- \rightarrow \gamma$, it violated the law of momentum conservation (which requires two photons). Nonetheless, both Jeans and Eddington took annihilation very seriously. Although Eddington vacillated between the two possibilities of subatomic energy production, by the late 1920s he leaned towards the annihilation hypothesis, which not only was far more energy-efficient but also promised a much longer lifetime of stars than the fusion alternative. Only in

1935 did he abandon proton–electron annihilation, and then without declaring it an impossible process. By that time, the fusion hypothesis received indirect support from experiments that possibly demonstrated the transformation of hydrogen to helium. By bombarding heavy hydrogen in the form of ammonium chloride ND_4Cl with deuterons, in 1934 Rutherford and his collaborators at the Cavendish Laboratory interpreted the product to be the hitherto unknown hydrogen isotope H-3 (tritium):

$$^{2}_{1}\mathrm{H} + ^{2}_{1}\mathrm{H} \rightarrow ^{4}_{2}\mathrm{He}^{*} \rightarrow ^{1}_{1}\mathrm{H} + ^{3}_{1}\mathrm{H}.$$

As an alternative, they suggested that the unstable He-4 nucleus might decay into a He-3 nucleus and a neutron. However, Rutherford did not relate the process to either stellar energy or element formation in stars.

Although the real neutron dates from 1932, Rutherford had introduced the name in 1920 for a tightly bound and supposedly stable proton–electron composite. The neutron hypothesis was well known and taken seriously throughout the 1920s, especially by Cambridge physicists. When James Chadwick announced his discovery of the neutron in 1932, he thought at first that he had found Rutherford's neutral particle. If the particle existed as a stable object, as believed by Rutherford and his group, it would squarely contradict the proton–electron annihilation hypothesis. This might have been a problem, but in fact it was not since no one noticed it. Eddington was undoubtedly aware of Rutherford's neutron, but apparently it did not occur to either him or others that it might be of astrophysical relevance.

Eddington, and with him most astrophysicists, believed that the heavier chemical elements were formed in the stars from lighter elements, meaning ultimately hydrogen. But according to Jeans, and also to the German chemist Walther Nernst, stellar processes proceeded from the complex to the simple and started with highly radioactive transuranic atoms that gradually decomposed into radiation and lighter atoms (for the Jeans or Jeans–Nernst hypothesis, see Section 2.5.1). Jeans' hypothesis, in part meant as an answer to the stellar energy problem, was rejected by most physicists and astronomers, who considered it contrived and without a grain of empirical justification. Among the critics were

Rutherford and Eddington, who dismissed it as nothing but a speculation. Responding to what he thought was Jeans' antievolutionary view, Eddington wrote: "Personally, when I contemplate the uranium nucleus consisting of an agglomeration of 238 protons and 146 electrons, I want to know how all these have been gathered together; surely it is an anti-evolutionary theory to postulate that this is the form in which matter first appeared" (Eddington, 1928). Of course, Jeans did not deny cosmic and stellar evolution, he only questioned if this kind of evolution were of the same kind as the simple-to-complex evolution known from biology.

Eddington was undoubtedly a pioneer of what came to be known as nuclear astrophysics, but he was reluctant when it came to developing the ideas of stellar energy production which he only discussed in a vague and uncommitted way. Without attempting to harvest the fruits of the seeds he had planted, he left the field to a new generation of nuclear physicists, apparently uninterested in their results. It is quite remarkable that he ignored the progress which in the 1930s finally led to an understanding of the mechanism of stellar energy. Nowhere in his publications from 1939 to 1946 did he mention the breakthrough theory of Hans Bethe.

3.2.4. NON-CONSERVATION OF STELLAR ENERGY?

During about 1930, many physicists believed that the new and highly successful quantum mechanics faced serious problem when confronted with phenomena originating in the atomic nucleus. Apparently irresolvable anomalies turned up, possibly indicating that the laws of fundamental physics had to be reconsidered. A few years later, it became clear that most of the anomalies were rooted in the erroneous but universally shared view that matter consisted solely of protons and electrons. An atomic nucleus with mass number A and atomic number Z was conceived as a dense system of A protons and $A–Z$ electrons. At the time, it was not recognized or even dimly perceived that this picture of the nucleus might be erroneous. The most discussed of the anomalies was the continuous shape of the beta decay spectrum, which seemed to be irreconcilable with the generally assumed two-particle process

$$(A, Z) \rightarrow (A, Z+1) + \beta^-.$$

The mother and daughter nuclei can only exist in discrete energy states, and thus, if an electron were simply expelled from the nucleus, the result would be a discrete and not a continuous spectrum.

Much worried, Bohr thought that the beta anomaly indicated that the principles of energy and momentum conservation were invalid within the atomic nucleus. If beta decay were not governed by energy conservation, the continuous spectrum would no longer be a mystery. Moreover, Bohr thought that the cure for the beta decay problem, namely energy non-conservation, might also be the cure for the problem of stellar energy production discussed by Eddington and other astronomers. In an unpublished manuscript of June 1929, he wrote:

> "We may be prepared for a disturbance of the energy balance also in certain large scale phenomena. ... Perhaps in consideration of this kind we can find an explanation of the energy source which according to present astrophysical theory is wanting in the sun, and the origin of which is generally sought in the transformation of elementary electric particles into radiation. ... A test of these consequences might perhaps be obtained from a closer analysis of astrophysical evidence regarding the evolution of stars, in the interior of which, as ordinarily assumed, the reversal of radioactive processes takes place on a large scale" (Bohr, 1986, p. 88).

Bohr apparently had in mind a process in which an atomic nucleus in the interior of a star captured an electron moving at thermal speed and emitted it at the much higher speed of a beta electron. He did not suggest any details of the hypothetical reverse beta process, which left the atomic nucleus unchanged and therefore did not result in element synthesis in the stars. Bohr's reference to the reverse beta process as one "ordinarily assumed" is puzzling, since it was not part of either the fusion process or the annihilation process, the only candidates for stellar energy production ordinarily considered.

The ever-critical Wolfgang Pauli read Bohr's manuscript and did not like it at all. In a letter to Bohr, he advised, "Let this note rest for a good long time and let the stars shine in peace!" (Bohr, 1985, p. 446). Bohr followed the advice and decided not to publish his manuscript, but he maintained his view for another three years. Pauli, a firm believer in symmetry and conservation principles, was convinced that the law of energy conservation was universally valid and that the atomic nucleus was no exception. At the end of 1930, he came up with his own attempt, entirely different from Bohr's, to solve the riddle of the continuous beta spectrum in accordance with energy conservation. Pauli famously suggested that the decay was a three-particle process involving also a very light "neutron" — or what soon would be known as a neutrino — residing in the nucleus. Pauli's alternative, in a slightly modernized formulation, was

$$(A, Z) \rightarrow (A, Z+1) + \beta^{-} + \overline{\nu},$$

where the symbol $\overline{\nu}$ represents an antineutrino. Although Bohr only discussed his heretical idea of energy non-conservation in two of his publications, it was well known through his oral presentations and extensive communications by letter. However, it made little impact in the physics community and even less in the community of astronomers and astrophysicists. Bohr had studied Eddington's *The Internal Constitution of the Stars*, but he was never seriously interested in astrophysics, a subject that was peripheral to the research done at his institute in Copenhagen. George Gamow, a frequent visitor at the institute, was fascinated by Bohr's hypothesis of energy non-conservation, which he promoted in *Atomic Nuclei and Nuclear Transformations*, a pioneering textbook published in 1931. However, fascinated as Gamow was, he chose to ignore Bohr's speculation concerning stellar energy.

Another of the Russian visitors in Copenhagen, Lev Landau, was clearly sympathetic to and, to some extent, inspired by Bohr's general idea that the interior of the stars required reconsideration of the fundamental laws of physics. In an important paper published in 1931, 22-year-old Landau argued that stars with masses greater than 1.5 solar masses must possess regions in which the laws of quantum mechanics fail. With respect to the energy emitted by the stars, he referred to the fact (as it was still believed to be) that electrons and protons live peacefully together in the atomic nucleus. According to Landau, this alone was enough to rule out proton–electron annihilation as a source of stellar energy. Fortunately, there was a better alternative:

> "Following a beautiful idea of Prof. Niels Bohr's we are able to believe that the stellar radiation is due simply to a violation of the law of energy, which law, as Bohr has first pointed out, is no longer valid in the relativistic quantum theory, when the laws of ordinary quantum mechanics break down (as it is experimentally proved by continuous-rays-spectra and also made probable by theoretical considerations). We expect that this must occur when the density of matter so great that atomic nuclei come in close contact, forming one gigantic nucleus" (Landau, 1932).

Shortly later, Landau and other physicists, among them Gamow and Pauli, realized that Bohr's hypothesis disagreed with arguments based on the general theory of relativity. This objection, together with the discovery of the neutron and the exclusion of electrons from the atomic nucleus, effectively put the idea of energy non-conservation in the grave. Bohr was still not completely convinced of his mistake, for the neutron was still controversial and besides, did Einstein's theory of general relativity apply to nuclear particles? All the same, by 1933 Bohr's "beautiful idea" was dead. The last time Bohr referred to energy non-conservation in the stars was at the 1933 Solvay congress, and then without defending the idea.

3.2.5. QUANTUM MECHANICS ENTERS THE GAME

The theories of stellar energy proposed by Eddington and Jeans were classical, in the sense that they made no use of quantum theory in either the old version or the new Heisenberg–Schrödinger version of 1925–1926. When quantum mechanics appeared, it was for a time uncertain if it also applied to the enigmatic atomic nucleus. That it did, or at least was useful in this exotic area of physics, was first demonstrated in an important paper by Gamow, who in August 1928 showed how the expulsion of alpha particles from the nucleus could be understood on the basis of

Schrödinger's wave mechanics. The same result was independently obtained one month later by the American physicists Edward Condon and Ronald Gurney. If quantum mechanics could explain the emission of alpha particles, perhaps it could also explain the inverse process, the entrance of a charged particle through the potential barrier of a nucleus. In that case, the theory might be astronomically relevant by suggesting how heavier elements were built up in the stars from simple elements and, as a result, produced a vast amount of stellar energy.

After discussions with Gamow and directly building on his theory, in the spring of 1929 a British and a German physicist, Robert d'Escourt Atkinson and Friedrich (or Fritz) Houtermans, coauthored an innovative paper that marked the beginning of nuclear astrophysics based on quantum mechanics. Atkinson was doing postgraduate work in Germany and had met Houtermans in Berlin. The title of their paper was "Zur Frage der Aufbaumöglichkeiten der Elemente in Sternen" (On the Question of Possible Syntheses of Elements in Stars), indicating that the focus was on element formation rather than stellar energy. The two young authors, aged 31 and 25, respectively, had originally titled their paper "Wie kann man ein Helium Kern in einen potential Topf köchen?" (How can one Cook a Helium Nucleus in a Potential Pot?), but the editor of *Zeitschrift für Physik* found it too imaginative for a respectable scientific journal.

Atkinson and Houtermans applied the inverse Gamow theory to the interior of a star with properties as assumed by Eddington, meaning a temperature of 40 million degrees and density of approximately 10 g/cm^3. Under these conditions, they found that the probability of alpha particles entering even a light nucleus with a low potential barrier was vanishingly small. For protons reacting with a target nucleus of atomic number Z, they derived an expression for the reaction probability W, technically known as the cross-section, of the form

$$W = \exp(-aZ/v), \qquad a = 4\pi e^2/h,$$

where v is the velocity of the incoming proton. Translating the velocity v to a temperature T by means of formulae from the Maxwell–Boltzmann gas theory, they concluded that W would increase with T and decrease with Z.

Assuming the mentioned values of temperature and density, Atkinson and Houtermans found that the average lifetime for protons reacting with helium (Z = 2) would be 8 seconds, whereas for neon (Z = 10) it would rise drastically to one billion years. The probability that a proton penetrated into the nucleus of a lead atom (Z = 82) came out as 10^{-69} or practically zero. The two physicists suggested that the main source of stellar energy might the process envisaged by Eddington, that is, a transformation of four protons and two electrons into a helium nucleus. However, contrary to the improbable six-particle collision, they imagined a cyclic process taking place by the consecutive capture of the six particles by a light nucleus and the subsequent expulsion of a helium nucleus in the form of an alpha particle.

The Atkinson–Houtermans theory was a promising beginning, but as the two authors realized, it was far from the final solution to the intertwined problems of energy production and element formation in the stars. Describing the formation of heavy elements as a "complete mystery," they ended their paper by confessing that it was only a partial success: "We believe that we have overcome the most important physical difficulties concerning the hypothesis of helium formation as the source of the stars' energy; but we have not dealt with some of the astrophysical paradoxes associated with this hypothesis" (Atkinson and Houtermans, 1929). And then, with a reference to the ideas of Jeans and Eddington, "We can in no way rule out the possibility that total annihilation of matter is the true energy source."

At a time when nuclear astrophysics was foreign to the large majority of astronomers and only of marginal interest to most physicists, the paper by Atkinson and Houtermans attracted but little immediate interest. It became more widely known only in 1931, when Atkinson published a more comprehensive and ambitious version of the theory in *Astrophysical Journal* and, to secure further dissemination, added a summary in *Nature*. Among other things, Atkinson investigated not only main sequence stars but also Cepheid variables, giants, and dwarf stars. Importantly, he took advantage of the recent recognition, only obtained about 1930, that hydrogen is by far the most abundant element in the Sun's atmosphere. Just five years earlier, it was generally accepted that the stars have approximately the same chemical composition as the

Earth, but in 1928–1930 works by Henry Norris Russell and others changed the picture completely. The new result suggested that hydrogen might also be the predominant element in the interior of the Sun and other stars, such as Atkinson assumed.

As Atkinson dimly perceived, the theory based on his collaboration with Houtermans was not limited to the formation of energy and elements in the stars but might also have cosmological consequences. Writing in the summer of 1931, shortly after the expanding universe of Friedman and Lemaître had replaced the earlier static models (Section 3.4.3), Atkinson suggested a scenario of how the present universe might have evolved from a primordial state:

> "It thus appears that as a result of the wave mechanics on one hand, and the general theory of relativity on the other, the universe may have developed its present complexity of stars and of atoms from an initial state consisting of a fairly dense, nearly uniform, nearly stationary mass of cold hydrogen. This comparatively simple beginning constitutes at least a pleasant ornament, if not an actual support for our theory" (Atkinson, 1931a).

Likewise, his paper in *Astrophysical Journal* dated March 1931 included the observation that "General relativity leads of course to the view that space as a whole, in which are imbedded the extragalactic systems, is expanding, and the motion is also observed, as is well-known" (Atkinson, 1931b). Atkinson's cosmological scenario had elements in common with Lemaître's theory and was prescient by appealing to both general relativity and quantum mechanics. But he did not follow-up, since it was after all just casual remarks. Still it is interesting that an astrophysicist not involved in cosmological models or extragalactic astronomy at this early date characterized the expanding universe as if it were generally accepted. It was only with Weizsäcker's later work (see the next section) and in particular with Gamow's work in the 1940s that Atkinson's scenario was developed from a speculation to a proper theory of the early universe.

Although the Atkinson–Houtermans theory and its development into Atkinson's more comprehensive theory of 1931 was an important step forwards, its impact was limited. The theory was still based on the presupposition that atomic nuclei contained electrons, and after the discovery of the neutron the theory was widely regarded as obsolete and no longer of interest. The neutron initiated a new chapter in the nascent field of nuclear astrophysics and led eventually, with Bethe's theory of 1939, to an understanding of the mechanism of stellar energy generation.

3.3

The Rise of Nuclear Astrophysics

3.3

The Rise of Nuclear Astrophysics

Although the first cautious applications of nuclear physics to areas of astronomy date from the early 1920s, it was only with the *annus mirabilis* of 1932 that nuclear astrophysics took off as a promising interdisciplinary branch of science. What made the year miraculous from an astrophysical point of view was the discovery of the neutron and the deuteron, and generally the recognition that atomic nuclei consists of protons and neutrons. Two lines of interconnected research programs followed, one focusing on the synthesis of elements in stars and the other on the nuclear processes responsible for their energy production. The early work on nuclear astrophysics was completely dominated by physicists, whereas astronomers and classically trained astrophysicists hesitated in entering into or even acknowledging the field. Thus, nuclear astrophysics was absent from the authoritative *Handbuch der Astrophysik* (Handbook of Astrophysics) published in eight volumes during 1929–1933. The subject only entered in a supplementary volume from 1936, and even then without giving it much space.

As far as the energy production in the Sun and other stars is concerned, the first and lasting success was undoubtedly Hans Bethe's monumental paper of 1939 in which he presented an almost full explanation of how nuclear reactions in the interior of the Sun results in an energy flux in convincing agreement with observations. While Bethe was not much interested in element synthesis, this was of deep concern to Fred Hoyle, who in 1953 came up with an ingenious explanation of how three alpha particles can unite to a carbon nucleus and from there form even heavier elements. The provisional climax of this line of research was a famous four-author paper of 1957 known colloquially as the B^2HF paper. The work on nucleosynthesis by Hoyle and others related to various kinds of stars, but indirectly it also had cosmological implications. If all elements could be cooked in stars, why consider a hot big bang as the hypothetical site of element formation?

At about the same time, astrophysicists began taking the neutrinos emitted by the Sun seriously. Although very difficult to detect, by the late 1960s experts knew the approximate value of the observed solar neutrino flux, which however disagreed significantly with the theoretically predicted value. Despite improved experiments and calculations, the "solar neutrino problem" persisted over the next couple of decades until it, eventually, at the beginning of the new millennium, became a challenging non-problem. The problem was solved in the sense that observation and theory came to an agreement, but the solution required revision of the standard model of elementary particles.

3.3.1. NUCLEAR ARCHEOLOGY

Since the mid-1930s, Gamow's research focused increasingly on astrophysics, an area which he, to a large extent, considered to be just nuclear physics applied to the stars. He was particularly fascinated by the prospect of explaining the relative abundances of the chemical elements by means of nuclear processes. This program was not Gamow's invention, for it was also part of Atkinson's earlier work (Section 3.2.5) and had been initiated as early as 1917 by the chemist William Harkins. Without knowing much of the atomic nucleus, Harkins was the first to suggest a link between geochemical and meteoric data on the one hand and nuclear physics on the other. He argued that the abundance distribution of elements in meteorites would be an important clue to the structure and formation of atomic nuclei.

Gamow had the great advantage over previous workers that he could make use of neutrons (n), deuterons (d), and positrons (e^+) as crucial agents in nuclear building-up processes. According to a theory Gamow first proposed in 1935 and revised a few years later, consecutive neutron capture was the basic mechanism in the formation of heavier elements from an original stellar state of pure hydrogen. The neutrons first had to be created, which Gamow suggested might occur through a two-stage process, the first of which was the production of deuterons from protons: $p + p \rightarrow d + e^+$. When enough deuterons had been produced, neutrons would appear as a result of the fusion process $d + d \rightarrow {}^3\mathrm{He} + n$. Neutron captures and subsequent beta decays would result in elements with a higher atomic number.

What Harkins, Atkinson, and Gamow were aiming at was not only to produce heavier elements in the stars, but also to do it in such a way that the result reflected the observed abundance of elements in the cosmos. This ambitious research program has aptly been called "nuclear archeology," namely to reconstruct the evolution of the universe by means of hypothetical stellar and cosmic nuclear processes and to test these by comparing the resulting pattern with the one known empirically. For the nuclear-archeological program to succeed, the abundances of elements and their isotopes had to be known with some degree of reliability, not only for the Earth and the solar system, but also for the stars. Such data did not exist in the early 1920s, when only the chemical composition of the Earth's crust was reasonably well known, whereas the distribution of elements in stars and galaxies was little more than guesswork. Reliable data only appeared in 1938, due to a comprehensive collection of data laboriously compiled by the Norwegian geochemist and mineralogist Victor Goldschmidt, professor of geology at the University of Oslo.

Goldschmidt's abundance data included not only elements in terrestrial and meteoric matter but also those detected in the atmospheres of the Sun and other stars. He demonstrated that, apart from considerable local variations, the abundance of elements roughly decreased with the atomic weight until about $A = 100$, after which it stayed approximately constant. By far, most of the universe was made up of hydrogen and helium, after which followed smaller amounts of carbon, nitrogen, oxygen, and neon. On the other hand, lithium, beryllium, and boron were abnormally rare, which suggested that the nuclei of these elements were particularly difficult to build up in stellar processes. Goldschmidt also studied the abundance of isotopes and their variation with the atomic number Z and the neutron number N. Nuclei with certain neutron numbers ($N = 30, 50, 82, 108$) had unusual abundances, which he thought reflected the processes by which complex nuclei had been formed from primary particles in the interior of stars and supernovae.

Although published in a relatively obscure journal, the proceedings of the Norwegian Academy of Science, Goldschmidt's tables quickly became known and recognized as crucial data for theories in astrophysics and cosmology aiming at understanding the formation of elements. Over a period of more than twenty years, the tables, either in their original form or in revised versions, functioned as an important data base in nuclear astrophysics and the nascent field of big-bang cosmology. For example, in 1942 Subrahmanyan Chandrasekhar and his student Louis Henrich at the Yerkes Observatory published a study of how elements might have been formed in a hot prestellar stage of the universe. Their aim was "to compare the computed theoretical abundance-curves with the observed relative abundances of the stable nuclei, as given by V. M. Goldschmidt" (Chandrasekhar and Henrich, 1942). Although the two authors found that "the theoretical abundance-curve under these conditions [$T = 8 \times 10^9$ K, $\rho = 10^7$ g/cm^3] agrees with Goldschmidt's curve quite satisfactorily for all elements from oxygen to sulphur," for heavier elements the predicted abundances disagreed completely with those observed.

Also the young German physicist Carl Friedrich von Weizsäcker relied on Goldschmidt's data in an important work of 1938, which qualifies as an early contribution to physical big-bang cosmology. Whereas in 1937 Weizsäcker had proposed to build up heavier elements by stellar processes starting with proton–proton reactions — essentially the same as those considered by Gamow, but supplemented with other neutron-producing reactions — in 1938, he argued that element formation probably took place before the stars had come into being. He now considered a state of the early universe characterized by the extreme temperature of about 10^{11}K and a no less extreme density in the neighborhood of the density of an atomic nucleus of the order 10^{14} g/cm^3. According to Weizsäcker's scenario, the gross distribution of the elements was obtained under these extreme conditions, whereas, subsequently, with the decrease of temperature and density, new nuclear reactions would occur and provide the fine distribution.

Contrary to his 1937 paper, Weizsäcker did not start with an environment consisting solely of hydrogen, which had the advantage that he could make use of other elements to propose a new cyclic model for the generation of nuclear energy. What became known as the CN cycle (or CNO cycle), a model which in a short time would be presented in greater detail by Bethe, presumed the catalytic action of carbon-12 nuclei. Without taking into regard the associated gamma quanta, Weizsäcker stated the cycle in this form:

$$^{12}\mathrm{C} + {}^{1}\mathrm{H} \rightarrow {}^{13}\mathrm{N},\ {}^{13}\mathrm{N} \rightarrow {}^{13}\mathrm{C} + e^{+}$$

$$^{13}\mathrm{C} + {}^{1}\mathrm{H} \rightarrow {}^{14}\mathrm{N}$$

$$^{14}\mathrm{N} + {}^{1}\mathrm{H} \rightarrow {}^{15}\mathrm{O},\ {}^{15}\mathrm{O} \rightarrow {}^{15}\mathrm{N} + e^{+}$$

$$^{15}\mathrm{N} + {}^{1}\mathrm{H} \rightarrow {}^{12}\mathrm{C} + {}^{4}\mathrm{He}.$$

Thus, the net process was 4H $\rightarrow$ He, the one originally contemplated by Harkins and Eddington. However, Weizsäcker did not calculate the energy release and he did not believe that the cycle, including the formation of the carbon nuclei, could take place in the interior of the stars.

The astrophysical parts of Weizsäcker's 1938 essay were interesting, but in retrospect the cosmological parts were even more interesting. Weizsäcker imagined a large mass of primeval matter, perhaps consisting of hydrogen and comprising the entire universe, which had once contracted gravitationally and formed the extreme conditions under which element formation would occur. He referred to Hubble's redshift measurements and the age of the expanding universe:

> "If one may interpret the redshift in the spectra of spiral nebulae as a Doppler effect, the backward extrapolation of this explosive motion provides a concrete reason to attribute a physical state significantly different from today's to the world of about 3×10^9 years ago. ... [This] leads to a new, independent age

> determination that corresponds well with known data. The radioactive elements still present today must, if they were not produced continuously in the stars, have originated at a time not significantly further back than their half-life" (Weizsäcker, 1938).

Weizsäcker further suggested that his speculation provided a cause for the cosmic expansion, namely the kinetic energy given by the original nuclear reactions. However, his scenario was no more than an occasional adventure into physical cosmology, a subject to which he did not return. Although Weizsäcker's picture of a universe originating in a hot superdense state had features in common with Lemaître's primeval-atom hypothesis (Section 4.1.2), he did not refer to either this model or other cosmological models based on the field equations of general relativity.

3.3.2. BETHE'S BREAKTHROUGH

It is generally accepted that the old riddle of solar energy was largely solved with two papers of 1938–1939 authored or coauthored by the German-American physicist Hans Bethe. Recognized as one of the brightest theoretical physicists of his generation, Bethe (whose mother was Jewish) fled Nazi Germany in 1933 and in 1935, at the age of 29, joined the faculty of Cornell University, New York. In 1941, he became naturalized as a citizen of the United States. Bethe was a great specialist in quantum mechanics and nuclear physics, and when he wrote his first paper on astrophysics in 1938, it was with this background and none in astronomy.

In March 1938, Bethe participated in a physics-astronomy conference on "Problems of Stellar Energy-Sources" in Washington, D.C. organized by Gamow and Edward Teller. There he met, among others, 27-year-old Charles Critchfield, a graduate student supervised by Gamow and Teller who was studying the energy liberated in nuclear processes. As a result, Bethe and Critchfield developed a quantitative theory based on the reaction $p + p \rightarrow d + e^+$ that had earlier been considered by Atkinson, Gamow, and Weizsäcker to be very rare and serving only to initiate other reactions. The detailed calculations of Bethe and Critchfield proved otherwise. The title of the two physicists' paper in *Physical Review* did not reveal that it was an important contribution to astrophysics: "The Formation of Deuterons by Proton Combination."

According to Bethe and Critchfield, a series of consecutive reactions resulted in the net process $4p \to {}^4\text{He} + 2e^+$. The proposed proton–proton or *pp* cycle started with the mentioned proton-to-deuteron reaction, after which the deuteron reacted with another proton according to $d + p \to {}^3\text{He} + \gamma$. It was followed by two further processes:

$${}^3\text{He} + {}^4\text{He} \to {}^7\text{Be} + \gamma \quad \text{followed by} \quad {}^7\text{Be} \to {}^7\text{Li} + e^+,$$

and

$${}^7\text{Li} + p \to {}^4\text{He} + {}^4\text{He}.$$

Calculations of the reaction probabilities and the energies involved resulted in a remarkably good agreement with observations. As Bethe and Critchfield summarized:

> "The energy evolution due to the reaction is about 2 ergs per gram per second under the conditions prevailing at the center of the sun (density 80, hydrogen content 35% by weight, temperature 2×10^7 degrees). This is almost but not quite sufficient to explain the observed average energy evolution of the sun (2 erg/g/sec)" (Bethe and Critchfield, 1938).

For the increase of the reaction rate with the temperature T, they found that it varied approximately as $T^{3.5}$. The reason why Bethe and Critchfield judged their prediction to be "almost but not quite sufficient" was that it presupposed the Sun's temperature and density to be constant over its volume. If the decrease of the two quantities toward the surface was taken into account, the result would be a considerably smaller energy production. For this reason, they concluded that "there must be another process contributing somewhat more to the energy evolution of the sun."

What this other process was became clear in a later issue of *Physical Review* including Bethe's celebrated article "Energy production in stars" in which he discussed the CN cycle in penetrating detail. This long and complex paper was submitted on 7 September

1938, less than a month after the appearance of the Bethe–Critchfield *pp* paper, but it only appeared in print in March 1939. The reason for the delay was that Bethe first submitted the manuscript for a prize award, which required that it was unpublished (he won the prize of 500 US dollars). In a qualitative sense, Bethe's version of the proposed CN cycle was the same as the one suggested by Weizsäcker, but in any other sense the two versions differed significantly. First of all, Bethe's theory was quantitative, including calculations of the probabilities of the individual reactions and making use, when available, of laboratory data for the evolved energies. It is also worth noting that Bethe was aware of the role played by the still hypothetical neutrinos in solar processes. The neutrino originally proposed by Pauli was at the time broadly accepted by physicists but rarely turned up in astrophysical contexts. Without distinguishing between neutrinos and antineutrinos Bethe discussed the process $p+\overline{\nu}\rightarrow n+e^{+}$, where $\overline{\nu}$ denotes an antineutrino. He calculated that, by far, most of the neutrinos would leave the Sun without causing nuclear reactions. As we shall see in Section 3.3.5, much later the question of solar neutrinos would become a hot topic in astrophysics.

Whereas Weizsäcker focused on the synthesis of elements, Bethe's focus was exclusively on energy production and he refrained from speculating on a primeval state of stellar matter. In other words, cosmology and cosmogony were absent from Bethe's paper. Only at the end of the paper did he briefly refer to the cosmological age paradox, noting that the lifespan of stars "is long compared with the age of the universe as deduced from the red-shift (~2 × 10^9 years)" (Bethe, 1939). The existence of carbon as a catalyst was a *sine qua non* for Bethe's theory, as it was for Weizsäcker's, but Bethe just assumed the presence of carbon nuclei. In fact, he argued, apparently convincingly, that carbon-12 could not possibly be built up from hydrogen. More generally, "there is no way in which nuclei heavier than helium can be produced permanently in the interior of stars under present conditions. We can therefore drop the discussion of the building up of elements entirely and can confine ourselves to the energy production which is, in fact, the only observable process in stars" (Bethe, 1939). This prediction of Bethe's turned out to be wrong.

The main result of the 1939 theory was the CN cycle, which Bethe argued was the principal energy source for main-sequence stars, including the Sun. Having calculated the temperature dependence of both cycles, approximately $T^{3.5}$ for *pp* and T^{16} for CN, he concluded that for temperatures below 16 million degrees the *pp* cycle would dominate, for higher temperatures the CN cycle. Turning the problem around, he found that the CN cycle, in order to give the energy produced by the Sun, would require a central temperature of 18.5 million degrees, in excellent agreement with the value of 19 million degrees based on astrophysical models of the Sun.

Bethe's theory of stellar energy became an instant success, universally hailed as a breakthrough in nuclear astrophysics. While John Herschel in 1833 had admitted that "we know nothing" about the Sun's production of heat (Chapter 3.2), with Bethe's theory 106 years later "nothing" became "almost everything." In 1967, Bethe was belatedly awarded the full Nobel Prize in physics for "his contributions to the theory of nuclear reactions, especially for his discoveries concerning energy production in stars." Incidentally, this was the first time ever that the Nobel Committee in Stockholm honored a contribution to astrophysics. Time was more than ripe for Bethe's prize, for until the year of his award he had been nominated no less than 48 times by a series of prominent physicists and astronomers, first in 1943. However, until 1967 the Nobel Committee hesitated in recognizing astrophysics as a proper branch of physics, such as the committee conceived the term.

Although a great success, of course Bethe's theory was more the beginning of a new chapter than the end of the story of why the stars shine. During the next decade or so, the relative importance of the *pp* cycle and the CN cycle was much discussed, with the result that by the early 1950s it became generally accepted that the crossover point between the two cycles was about 1.5 solar masses. Thus, the Sun is mainly powered by the *pp* cycle and not by the CN cycle, as Bethe had originally concluded. According to present knowledge, the core temperature of the Sun is 15.7 million degrees and 90% of its emitted energy is due to the *pp* cycle, with the remaining 10% due to the CN cycle. As the Sun grows older, the core temperature will increase and so will the contribution from the CN cycle.

3.3.3. THE TRIPLE ALPHA PROCESS

Although Bethe was not concerned with nucleosynthesis — the building up of heavier elements from lighter ones — in his paper of 1939 he did discuss what he called the triple collision of alpha particles, that is, the direct formation of carbon-12 by the process $3\alpha \rightarrow {}^{12}C$ (where $\alpha = {}^{4}He$). However, he found that the yield of carbon would be negligible unless the temperature was about one billion degrees, much higher than the temperature in the interior of the Sun and similar stars. Yet, somehow carbon, oxygen and the other elements had come into existence. The question of how this had happened was not only of importance to astrophysics but also to the new field of big-bang cosmology as developed in the late 1940s by Gamow and his collaborators Ralph Alpher and Robert Herman. Gamow and his associates believed that all elements had been formed shortly after the big bang but were unable to find nuclear reactions that, under the temperature and density conditions of the early universe, led to elements heavier than helium. As they were forced to realize, there were no stable nuclei of mass 5 or 8 that could serve as "bridges" between helium and carbon. In spite of many attempts, the so-called mass-gap problem remained unsolved.

In 1952, the Austrian–American physicist Edwin Salpeter argued that in red-giant stars at $T > 10^8$ K, three alpha particles could fuse into carbon, not in a direct reaction but mediated by a small amount of the highly unstable beryllium-8 isotope known to have a lifetime of only 10^{-16} s. The calculated rate of reaction was very small, but at $T = 2 \times 10^8$ K Salpeter found that it was just high enough that a beryllium-8 nucleus, before decaying into two alpha particles, could absorb another alpha particle and form carbon-12. The reactions were

$${}^{4}He + {}^{4}He \rightarrow {}^{8}Be + \gamma \quad \text{and} \quad {}^{8}Be + {}^{4}He \rightarrow {}^{12}C^{*} \rightarrow {}^{12}C + \gamma,$$

where the asterisk marks a short-lived, highly unstable state of the carbon atom. Salpeter further suggested that, once carbon had been produced, subsequent capture of additional alpha particles might result in still heavier elements, for example by ${}^{12}C + \alpha \rightarrow {}^{16}O$ and ${}^{16}O + \alpha \rightarrow {}^{20}Ne$. Nuclear processes similar to Salpeter's had been proposed a little earlier by Ernst Öpik, an astronomer at the Armagh Observatory in Ireland, but his paper in the *Proceedings of the Royal Irish Academy* attracted almost no attention. When Salpeter wrote his paper on the triple-alpha process, he was unaware of Öpik's work.

Fred Hoyle was deeply concerned with problems of nucleosynthesis even before he cofounded the steady-state theory of the universe in 1948. He spent the first three months of 1953 at Caltech, where William Fowler and his colleagues at the Kellogg Radiation Laboratory specialized in experimental nuclear physics related to stellar processes. Hoyle knew about Salpeter's theory of the triple alpha process but thought that its predicted reaction rate was too small to generate carbon in red giants of a temperature close to 10^8 K. He consequently decided to reconsider the details of the theory. As he recalled:

> "Salpeter's publication of the 3α process freed me to take a fresh look at the carbon production problem. I found difficulty in generating enough carbon, because the carbon kept slipping away into oxygen as it was produced. A theoretically possible way around this difficulty was greatly to speed-up the carbon synthesis by a rather precisely tuned resonance which would need to be about 7.65 MeV above ground-level in the ^{12}C nucleus" (Hoyle, 1986).

That is, Hoyle realized that to get an appreciable fraction of the original helium transformed into carbon, there had to be in the carbon nucleus an excited energy level, a so-called resonance state, exactly of the mentioned value (Hoyle's value in 1953 was actually 7.68 and not 7.65 MeV). With the new hypothesized resonance, the carbon yield would increase drastically, with a factor of about 10 million times the one characterizing the Salpeter process. The predicted resonance allowed Hoyle to estimate the abundance ratio between carbon-12 and oxygen-16 to be about 1:3, in good agreement with spectroscopic evidence. But was the 7.68 MeV resonance more than just a theoretical deduction? Did it really exist?

Hoyle approached Fowler and the other experimentalists at the Kellogg Laboratory and persuaded them to look for the missing resonance. Within a few weeks, Ward Whaling, a physicist at the laboratory, found the carbon resonance at the level 7.68 ± 0.03 MeV, in excellent agreement with Hoyle's calculations. Further experiments by Fowler and his team narrowed down the resonance level to 7.653 ± 0.008 MeV. Thus, Hoyle's theoretical work was beautifully confirmed. At the time, the discovery of what modern astrophysicists refer to as the "Hoyle state" was considered interesting but not of great importance. Not even Hoyle saw it as particularly important. That only changed many years later, when the prediction was associated with the existence of life. Carbon is of course a prerequisite for all life as we know it, and without Hoyle's resonance there would be no or only very little carbon, hence no life in the universe. We would not exist! As two cosmologists phrased it in a book of 1986: "Hoyle realized that this remarkable chain of coincidences — the unusual stability of beryllium, the existence of an advantageous resonance level in C^{12} and the non-existence of a disadvantageous level in O^{16} — were necessary ... conditions for our own existence and indeed the existence of any carbon-based life in the Universe" (Barrow and Tipler, 1986, p. 253).

The so-called anthropic principle was named and first formulated by the Australian-British astrophysicist Brandon Carter in a lecture of 1973. The principle or assumption exists in several versions, some more speculative than others, but essentially it is the claim that the laws of nature and the evolution of the universe must be compatible with our existence (or with the existence of complex organic life in general). Ever since Carter's announcement, the anthropic principle has been highly controversial. Does it belong to science, as Carter argued, or should it rather be considered a philosophical claim? Are the kind of explanations offered by the anthropic principle scientifically acceptable, or are they merely pseudo-explanations? If of scientific value, the principle must result in predictions of the same kind as known from other theories, preferably in precise predictions of new phenomena. It is very hard to find confirmed anthropic predictions of this kind, except that Hoyle's prediction of the carbon resonance has repeatedly been claimed to be a case. However, there is little doubt that all such claims are unjustified.

There is no evidence whatsoever that Hoyle thought of life or anthropic fine-tuning when he came up with his prediction in 1953, and nor did other scientists at the time make a connection to life.

Retrospectively, the prediction can be interpreted as anthropic, but this is irrelevant and is to read history backwards. Nonetheless, late in life, Hoyle came to the conclusion that his 1953 prediction could be understood as anthropic in a *post factum* sense:

> "It was shown in 1952–1953 that to understand how carbon and oxygen could be produced in approximately equal abundances, as they are in living systems, it was necessary for the nucleus of ^{12}C to possess an excited state close to 7.65 MeV above ground level. No such state was known at the time of this deduction but a state at almost exactly the predicted excitation was found shortly thereafter. So one could say that this was an example of using the weak anthropic principle in order to deduce the way the world must be" (Hoyle and Wickramasinghe, 1999).

On the other hand, Hoyle never suggested that his motivations for the prediction were related to the existence of life.

3.3.4. STELLAR NUCLEOSYNTHESIS

In early papers from 1946–1948, Hoyle investigated the formation of heavy elements in stars by means of thermonuclear reactions taking place at very high temperatures. He suggested that the necessary temperatures and densities indicated supernovae as the sources of element formation but without proposing definite nuclear processes. This line of inquiry was further developed in a theory Hoyle completed during his stay at Caltech in 1953 and in which he elaborated on his recent theory of the triple alpha process. The main result of Hoyle's new theory, which appeared in print in 1954, was that the elements from carbon to nickel ($Z = 6 - 28$) were synthesized inside special types of collapsed stars with a temperature of the order of 100 million degrees and thus much hotter than the Sun.

After having returned to England, Hoyle joined forces with Fowler and Margaret and Geoffrey Burbidge, a married couple of British astrophysicists, to develop an even more comprehensive theory of how the elements had come into being. Ambitiously, the aim of the theory was to explain, at least in outline, the formation and distribution of *all* elements except hydrogen in the periodic system. The culmination of the British–American collaboration was a monumental paper on element synthesis published in 1957 in *Reviews of Modern Physics*. This comprehensive and complex 104-page article soon became recognized as a landmark contribution to nucleosynthesis, often referred to as the B^2HF work after the initials of the four authors.

The B^2HF paper considered eight different types of nuclear reactions, including the slow neutron-capture "*s*-process" and the very rapid "*r*-process" supposed to take place in supernovae. The names seem to have been coined by the four authors of the paper. While the *s*-process synthesized many of the elements up to bismuth ($Z = 83$), the even heavier elements were the result of *r* processes. Guided by new empirical abundance data more detailed than those compiled by Goldschmidt in the late 1930s (Section 3.3.1), Hoyle and his coauthors were able to account for the origin and abundance of almost all elements and their isotopes. In their discussion of the formation of the heaviest elements, they included the isotopes of the recently discovered transuranic elements such as californium, einsteinium, and fermium. The four authors proudly concluded that, "We have found it possible to explain, in a general way, the abundances of practically all the isotopes of the elements from hydrogen to uranium by synthesis in stars and supernovae" (Burbidge *et al.*, 1957). Hydrogen nuclei were outside the scope of the theory, since the stable protons were taken to be the original particles out of which the other elements were formed.

Although the B^2HF theory was explicitly astrophysical and not cosmological, there were a few indirect references in the paper to the ongoing controversy between the big-bang theory and the steady-state theory (the term "steady state" only occurred once and "big bang" not at all). Hoyle, Fowler, and the Burbidges emphasized that their theory had the methodological advantage that it built on nuclear processes currently taking place in the stars. The changing structure of stars during their evolution offered a great variety of processes available for the synthesis of different nuclear species. This was quite different from the "primeval theory," their name for Gamow's big-bang theory of the early universe. "The theory of primeval synthesis," they wrote, "demands that all the varying conditions occur in the first few minutes, and it appears highly improbable that it can reproduce the abundances of those isotopes which are built on a long time-scale in a stellar synthesis theory." For the controversy over the steady-state theory, see also Section 4.1.4.

In another passage the authors mentioned that "our conclusions will be equally valid for a primordial synthesis in which the initial and later evolving conditions of temperature and density are similar to those found in the interior of stars." However, they knew perfectly well that the conditions of the early Gamow universe differed greatly from those in the stars, and so the admission was vacuous. Gamow was forced to admit the strength of the B^2HF theory, and yet he disliked it, not only because it seemed contrary to his preferred cosmological theory but also because it represented a style of physics so different from his own. Confusingly complex and multifarious as the theory was, it lacked the elements of unity and simplicity that Gamow valued and which he found much better represented in his own picture of the exploding universe. One of his characteristically non-academic responses to the B^2HF theory was this:

> "What … Hoyle demands sounds like the request of an inexperienced housewife who wanted three electric ovens for cooking a dinner: one for the turkey, one for the potatoes, and one for the pie. Such an assumption of heterogeneous cooking conditions, adjusted to give the correct amounts of light, medium-weight, and heavy elements, would completely destroy the simple picture of atom-making by introducing a complicated array of specially designed 'cooking facilities'" (Gamow, 1952, p. 56).

Although the B^2HF theory was sometimes considered a "tremendous triumph" for steady-state cosmology (such as Hermann Bondi called it), it did not seriously weaken the big-bang alterative, which after all had two nuclear ovens at its disposal, the explosive beginning of the universe *and* the stars. Moreover, the generally successful B^2HF theory failed to account for the formation and abundance of helium, which, apart from hydrogen, was known to be the most common element in the universe. While sufficient amounts of helium could not be cooked in either ordinary or extraordinary stars, big-bang calculations resulted in an abundance of the same order as observed. By the mid-1960s, the helium problem, and also the related deuterium problem, turned out to be one more nail in the coffin of the now seriously weakened steady-state theory.

The high appreciation of the B^2HF theory is underlined by the fact that the four authors were nominated for the 1965 Nobel Prize. The nominator was the American physical chemist Harold Urey, who in the 1950s worked on the chemical composition of the cosmos and was himself a Nobel laureate of 1933 for his discovery of deuterium. Although nothing came out of this early nomination, in 1983 Fowler received half the physics prize for his work on "the nuclear reactions of importance in the formation of the chemical elements in the universe." The pioneering B^2HF theory was a major reason for the decision to honor him. It was widely felt in the astrophysical community that Hoyle ought to have been included in the prize, a feeling that Fowler shared. In his Nobel Prize Lecture, Fowler stressed that his own work was a continuation of Hoyle's, stating that "Although Bethe in 1939 and others still earlier had previously discussed energy generation by nuclear processes in stars the grand concept of nucleosynthesis in stars was first definitely established by Fred Hoyle" (Fowler, 1984).

Hoyle's missing Nobel Prize was to some extent compensated for when he, together with Salpeter, received the 1997 Crafoord Prize for "their pioneering contributions to the study of nuclear processes in stars and stellar evolution." Like the Nobel Prize, the Crafoord Prize, established in 1980 and of value half a million US dollars, is awarded by the Royal Swedish Academy of Sciences.

3.3.5. SOLAR NEUTRINOS

Pauli's elusive neutrino was first detected, and in this respect turned into a real particle, in experiments of 1956. Using as a source the beta decays from a nuclear reactor, what Frederic Reines and Clyde Cowan discovered was actually the antineutrino $\bar{\nu}$ and not the neutrino ν. It had been known since Bethe's 1939 theory that neutrinos are emitted from the Sun in large numbers as a result of the burning of hydrogen to helium (Section 3.3.2), and in the 1950s astrophysicists began to realize that measurements of solar neutrinos might yield important information of what happens in the interior of the Sun. Contrary to other particles, the very weakly interacting neutrinos pass largely unimpeded through the dense solar core with the speed of light. As the American physicist John Bahcall, a leading expert in neutrino astrophysics, put it:

> "No *direct* evidence for the existence of nuclear reactions in the interior of stars has yet been obtained ... Only neutrinos, with their extremely small interaction cross sections, can enable us to *see into the interior of a star* and thus verify directly the hypothesis of nuclear energy generation in stars" (Bahcall, 1964).

The first experiments to measure the solar neutrino flux were proposed by Ray Davis, a physicist at the Brookhaven National Laboratory, who in 1955, before the Reines–Cowan discovery, built a remarkable detector to measure the particles. The chlorine atoms in a large tank filled with carbon tetrachloride (CCl_4) would react with incoming neutrinos according to

$$^{37}\mathrm{Cl} + \nu \rightarrow {}^{37}\mathrm{Ar} + e^{-}.$$

The number of produced radioactive argon-37 nuclei could be measured radiochemically by their decay back in chlorine-37 by electron capture. However, Davis' early experiments failed to result in an estimate of the neutrino flux, and the situation only improved about a decade later and then only slowly.

In a much improved series of experiments, Davis and his team used as a detector a 100,000 gallon (390,000 liter) tank of perchloroethylene (C_2Cl_4) located some 1600 m underground in the Homestake gold mine in South Dakota to screen the detector from cosmic rays and other noise signals. The first measurements of 1968 indicated that less than 0.2 argon-37 atoms were produced per day, a neutrino flux due the decay of boron-8 in the Sun's interior: $^{8}\mathrm{B} \rightarrow {}^{8}\mathrm{Be} + e^{-} + \bar{\nu}$. Later experiments using gallium instead of chlorine as detector material measured the more high-energy neutrinos coming from the fusion process $p + p \rightarrow {}^{2}\mathrm{H} + e^{+} + \nu$. What matters is that the data from these experiments suggested a neutrino flux significantly lower than the one expected from the best theoretical solar model.

The result was disturbing and grew even more disturbing when it was realized that the discrepancy persisted even though experiments greatly improved and more advanced solar models were developed by Bahcall and other theorists. By 1970 the "solar neutrino problem" had been established. Despite numerous, ever-more accurate experiments and calculations over the next two decades, the problem would not go away. As a unit for the neutrino flux, physicists use a "solar neutrino unit" or SNU (pronounced "snew"), which corresponds to one neutrino being absorbed per second per 10^{36} chlorine atoms. The definition of the unit indicates how weakly interacting neutrinos are. Although one SNU is a very small number, fortunately Avogadro's number 6×10^{23} is very large. The number of chlorine atoms in the Homestake experiment amounted to nearly 10^{31}. In terms of the SNU unit, Bahcall reported in 1989 a predicted flux of 7.9 ± 2.6 as compared to the observed flux 2.1 ± 0.9. Obviously, something was seriously wrong.

The robustness of the solar neutrino anomaly was underlined by several large-scale experiments using methods and detectors very different from those of Davis' chlorine experiments or those of the gallium experiments. Whatever methods and detector technologies used, they all confirmed the puzzling deficit of neutrinos emitted from the Sun. For example, the advanced and very expensive Japanese Kamiokande II experiment found in 1990 a ratio between theory and observation of approximately one-half. In this experiment, electronic counters detected the rare interactions of neutrinos with electrons in an enormous volume of pure water, a method entirely different from earlier experiments. The even more advanced and expensive Super-Kamiokande, including 50,000 tons of water and 13,000 photomultiplier tubes, reported after more than three years of observation a measured flux of 2.32 ± 0.03 SNU, again incompatible with calculations based on the standard solar model.

It would seem that there were basically two ways of solving the problem, either by suspecting errors in the experiments or in the standard model of the Sun. However, given the variety and precision of the many experiments it was unlikely that they had some unrecognized experimental error in common. Nor was it likely that there could be a major error in the highly successful model of the Sun. There were a couple of other alternatives, such that some of the neutrinos might decay on their way from the Sun to the Earth. The hypothesis of decaying neutrinos was proposed but only to be shot down by observations.

Yet another and more promising possibility related to the fact that all solar neutrinos detected in the experiments were the ordinary neutrinos associated with the electron. It had been known since the early 1960s that there exists a separate neutrino associated with the heavier muon and that ν_μ is different from ν_e (there is also a third species, the tau neutrino ν_τ). By the late 1980s, the muon neutrino was well known, but since it does not occur in the nuclear reactions in the interior of the Sun, it was not expected to turn up in the solar neutrino flux. But perhaps the two kinds of neutrinos could change into one another as they traveled from the Sun to the Earth? Among the first to propose "neutrino oscillations" as a solution to the solar neutrino problem were Bahcall and Bethe, who in a joint paper of 1990 suggested that "the explanation of the solar-neutrino problem probably requires physics beyond the standard electroweak model with zero neutrino masses" (Bahcall and Bethe, 1990). More than half a century after having solved the riddle of the Sun's production of energy 84-year-old Bethe was ready to solve another astrophysical riddle.

The hypothesis was that the missing solar neutrinos were electron neutrinos that had changed into muon or tau neutrinos during their journey to the Earth and therefore had not been detected. Preliminary evidence for neutrino oscillations turned up at the Super-Kamiokande experiments, but it was only in the early years of the new millennium that electron and muon neutrinos from the Sun were directly observed and neutrino oscillations confirmed. The site was a new institution near Ontario called the Sudbury Neutrino Observatory that used heavy water as detecting material for neutrinos of all types. With data from the Super-Kamiokande and the Sudbury Neutrino Observatory, the solar neutrino problem disappeared. It turned out that the total number of neutrinos (e, μ, τ) agreed with the predicted number and that the electron neutrinos made up about one-third of the total number of observed neutrinos.

Understandably, the solution of the thirty-year old neutrino problem was hailed as a great triumph of nuclear astrophysics. The Nobel Prize for 2015 was awarded to the leaders of the two collaborations responsible for the discovery of neutrino oscillations, Takaaki Kajita (Super-Kamiokande) and Arthur McDonald (Sudbury Neutrino Observatory). While the neutrino had traditionally been thought to be massless, with the discovery of neutrino oscillations followed that it had to be a massive particle, albeit with a very small mass. The current upper limit for the neutrino mass is 1.1 eV, which is about 500,000 times less than the mass of an electron, which is 511 keV. Neutrinos weigh next to nothing, but on the other hand, there is an awful lot of them. In fact, the neutrino is the second-most abundant particle in the universe, only surpassed by the massless photon. Cosmologists estimate that in each cubic centimeter of the universe there are about 300 neutrinos, approximately 100 million times as many as there are hydrogen atoms.

Many of the crucial ideas leading to a solution of the solar neutrino problem were predicted by the Italian–Russian nuclear physicist Bruno Pontecorvo, a former collaborator of Fermi who in 1950 defected to the Soviet Union for political reasons. As early as 1945, Pontecorvo suggested that neutrinos might be detected by the chlorine-37 reaction later used by Davis, and in 1959 he realized that the still undetected muon-neutrino must be different from the electron neutrino. At about the same time, he came up with the idea that the ordinary electron neutrino may convert into other types of neutrinos, such as was eventually discovered in the form of neutrino oscillations. Pontecorvo died in Russia in 1993, too early to experience the final solution of the solar neutrino problem and the vindication of his brilliant predictions.

.3.4

An Expanding Universe

3.4

An Expanding Universe

Since about 1930, it has been known that the universe is not static but in a state of expansion. This momentous and unanticipated insight was a combination of theory and observation, where theoretical analysis based on the theory of general relativity was more important than the supporting evidence in the form of galactic redshifts. When astronomers and physicists were first confronted with the surprising claim of an expanding universe, many of them responded with confusion and disbelief. To an even higher degree, this was also the typical reaction of the lay public, who found the concept strange or plainly incomprehensible. In February 1931, Eddington introduced what is still today a much used pedagogical picture, the balloon analogy. Two years later, in his brilliant popular exposition *The Expanding Universe*, he presented the analogy to lay readers. "Let us represent spherical space by a rubber balloon," Eddington wrote. "Imagine the galaxies to be embedded in the rubber. Now let the balloon be steadily inflated. That's the expanding universe" (Eddington, 1933, p. 66). The analogy addresses the natural but illegitimate question of what the universe expands into. But what if the universe — or the balloon — is not spherical, but flat and infinite? If something expands, it becomes bigger, as does the surface of the balloon; but how can the infinite universe grow bigger?

There are other problems or at least other misconceptions. Among the latter are that if we trace a continually expanding universe backwards in time, we must necessarily end in a big bang. But there is no necessary connection between the two concepts, neither logically nor historically. Whereas the big-bang universe must be expanding, or at least have expanded in the past, an expanding universe does not need to have started in a big bang. If this were the case, one would expect that astronomers accepted the big-bang picture as a consequence of the expanding universe, which they did not. For example, Eddington and de Sitter were pioneers of and firm believers in the expanding universe, and yet they and most of their contemporaries in the 1930s denied that the universe had an origin in a big bang.

Then there is the question of *what* expands. Space blows up and the distances between galaxies and galactic clusters increases, but what about the distances between other structures? For a while, it was discussed even among experts in cosmology if the Milky Way grows bigger as a result of the cosmic expansion. It took a year or two until it was understood that it does not and that the gravitational forces within the Milky Way annul the expansive force. Likewise, objects which are held together by electromagnetic and nuclear forces remain at a fixed physical size as the universe swells around them. There were other questions about the expanding universe that were asked in the 1930s and which still puzzle a lay audience. To mention but one, how can a completely empty universe expand?

3.4.1. RELATIVISTIC COSMOLOGY

Insofar as theory is concerned, cosmology experienced a genuine revolution in February 1917 when Einstein published a treatise with the title "Kosmologische Betrachtungen zur allgemeine Relativitätstheorie" (Cosmological Considerations on the General Theory of Relativity) in the Proceedings of the Prussian Academy of Science. The world or universe that Einstein dealt with was conceptually different from what astronomers traditionally associated with the term, namely the stars and other celestial objects within the limits of observation. With Einstein's innovation, it became everything, the totality of events in space and time, including space-time itself as the fundamental substratum of the world. And all this was governed by a single tensor equation.

The formal foundation of Einstein's theory was the cosmological field equations, which in a slightly modernized formulation can be written as

$$G_{\mu\nu} - \Lambda g_{\mu\nu} = \kappa T_{\mu\nu}.$$

The indices μ and ν can attain four values corresponding to the four-dimensional space-time. The quantity to the left $G_{\mu\nu}$ is known as the Einstein tensor and expresses the geometry of space-time, whereas the quantity to the right $T_{\mu\nu}$ is the energy-momentum tensor with the gravitational constant ($\kappa = 8\pi G/c^2$) in front of it; $g_{\mu\nu}$ is the so-called metric tensor which describes the space-time continuum. For the cosmological constant Λ, Einstein found that it had to have a precisely fixed value, namely

$$\Lambda = \frac{4\pi G}{c^2}\rho,$$

where ρ denotes the average density of matter (see also Section 3.1.5). Contrary to Einstein's expectation, later the same year Willem de Sitter found another solution to the cosmological field equations corresponding to an empty but nonetheless spatially closed universe given by $G_{\mu\nu} = \Lambda g_{\mu\nu}$. For the radius of curvature, de Sitter derived the relation $R^2 = 3c^2/\Lambda$. Although Einstein was forced to accept de Sitter's alternative, he considered it to be artificial and unphysical since the curvature of space was given solely by the cosmological constant. During the period from 1917 to about 1930, there existed two model universes based on general relativity, one due to Einstein and the other to de Sitter. The two scientists agreed that there were no more solutions to the cosmological field equations. They also agreed that the universe was roughly the same in the indefinite past and would remain so in the indefinite future.

Although de Sitter's theory, being devoid of matter, might seem a very artificial candidate for the real world, it attracted considerable attention among mathematically minded astronomers and physicists. The lack of matter was a serious problem but not a devastating one, as it could be argued that the low matter density of the universe, of the order of 10^{-30} g/cm^3, made it almost empty. According to de Sitter, the density was so low that his model might apply as a zero-density approximation. An important reason for taking the de Sitter model seriously was that it (contrary to Einstein's model) predicted that clocks would run more slowly the farther away they were from the observer. Since frequencies are inverse time intervals, light would therefore be expected to be received with a smaller frequency or larger wavelength, that is, being redshifted. The "de Sitter effect" was seen as interesting because it promised a connection to the galactic redshifts observed a few years earlier. However, the redshifts appearing in the theory were an effect of a particular space-time metric and not a Doppler effect due to motion; neither was it due to an expansion of space. The de Sitter model would later be interpreted as representing an exponentially expanding universe, but to de Sitter and his contemporaries in the 1920s it represented a static universe.

Eddington, one of the most active contributors to the new relativistic cosmology, thought highly of both the Einstein model and the alternative de Sitter model. With a reference to the measurements of galactic redshifts, he wrote in 1923 that "It is sometimes urged against De Sitter's world that it becomes non-statical as soon as any matter is inserted in it. But this property is perhaps rather in favour of De Sitter's theory than against it" (Eddington, 1923, p. 161). On the other hand, a matter-filled universe was essential to Eddington's thinking of the universe, so he and some other cosmologists felt towards the end of the 1920s that neither the Einstein solution nor the de Sitter solution could represent the real universe. If the two models were the only possible solutions — and this was generally believed — how could cosmology still be based on general relativity?

Faced with this dilemma, a few scientists proposed models combining features of the two classical models. Some of these models were non-static in the sense that the metric depended on the time parameter, but from a physical point of view they nonetheless described a static and not an evolving universe. What later came to be seen as the obvious answer to the dilemma, namely to look for evolutionary models as solutions to the field equations, was only seriously considered by two somewhat peripheral scientists, and until 1930 their works were either ignored or unknown.

From a modern perspective, cosmology is a science solidly based on Einstein's field equations. It may be tempting to believe that this was also the case in the past, that astronomers quickly and enthusiastically recognized the Einsteinian revolution. But this was far from the case. To mention but one example, in the spring of 1924 the Swedish astronomer Carl Charlier gave a series of lectures on cosmology at the University of California which were published under the title "On the Structure of the Universe." Charlier's lectures were in the style of traditional stellar cosmology, his universe consisting of an infinite number of galaxies arranged in such a way that the matter density decreased with the distance of the observer. Nowhere did he refer to the general theory of relativity, and he also ignored the galactic redshifts discovered about a decade earlier. His only, indirect reference to the theories of Einstein, de Sitter and Eddington was this:

> "You know that there are also speculative men in our time who put the question whether space itself is finite or not, whether space is Euclidean or curved (an elliptic or hyperbolic space). Such speculations lie within the domain of possibility, but are of the same nature as the philosophical speculations ... in Kant's 'Kritik der reinen Vernunft.' They must be discussed on the support of facts and at present they must be considered as lying outside such a discussion" (Charlier, 1925).

By the mid-1920s, the large majority of astronomers shared Charlier's view, believing that relativistic cosmology was more a mathematicians' game than a respectable theory of the real universe as observed by means of telescopes. It was something that astronomers could safely ignore, which most did.

3.4.2. FROM SLIPHER TO HUBBLE

The discovery of spiral nebulae in the mid-nineteenth century initiated an era in which the nebulae (or galaxies) moved to the forefront of astronomical research. The enigmatic nature of the nebulae was seen as closely connected to another cosmic enigma, of whether the nebulae were grand structures similar to the Milky Way or much smaller objects within it. When Vesto Melvin Slipher, an astronomer at the Lowell Observatory in Arizona, started his observations of spiral nebulae in about 1910, most astronomers were in favor of the latter view. Using a specially designed spectrograph, in 1912 Slipher obtained spectra of the Andromeda Nebula (M31) which allowed him to conclude that the spectrum was shifted towards the blue corresponding to a motion towards the Sun at about 300 km/s. Five years later, he reported that he had found spectral shifts for 25 spirals and that most of them were redshifts rather than blueshifts. By 1925, he had collected data for 45 nebulae, of which 41 were redshifts. For one of the nebulae, he reported a redshift corresponding to a recessional velocity of no less than 1800 km/s. It seemed that recession was the rule to be found among the spiral nebulae.

Slipher and other astronomers translated spectral shifts to radial motion by means of a law first established by the Austrian physicist Christian Doppler in 1842. According to Doppler, if a light source moves relative to the observer with a radial velocity v, there will be a change in wavelength given by

$$z = \frac{\Delta\lambda}{\lambda} = \frac{\lambda' - \lambda}{\lambda} = \frac{v}{c},$$

where λ' is the measured wavelength and λ the one emitted by the source. The discovery of galactic redshifts attracted attention at an early date and was generally seen as evidence in favor of the "island universe" according to which the nebulae were huge structures similar to but far away from the Milky Way.

De Sitter was the first to consider the significance of the spectral shifts within the framework of relativistic cosmology, which he did in his theory of 1917, where he discussed only three nebulae of which one was the Andromeda with a blue-shifted spectrum. He speculated that the redshifts appearing in his own theory might be related to Slipher's discovery. De Sitter suggested that a systematic recession of the spiral nebulae would provide evidence in support of his own cosmological model, whereas it would be inexplicable according to Einstein's alternative model. Eddington too was intrigued by the cosmological meaning of the nebular redshifts but uncertain about how to interpret them. Referring to the receding nebulae, he wrote, "This may be a genuine phenomenon in the evolution of the material universe; but it is also possible that the interpretation of spectral displacement as a receding velocity is erroneous; and the effect is really the slowing down of atomic vibrations predicted by de Sitter's theory" (Eddington, 1920a, p. 161). In his widely read monograph, *The Mathematical Theory of Relativity* from 1923, Eddington included a list of 37 redshifts obtained from Slipher. He tended to believe that the redshifts were of deep cosmological significance, but at the time he was unable to figure out how to understand them within the context of general relativity.

With the increased number of nebular redshifts, it became natural to look for a definite relationship between the redshifts and the distances of the nebulae. Several such relations of the type $z = z(r)$ were proposed on theoretical grounds, but none of them could be justified observationally simply because the distances to the spirals were still unknown. The situation only changed after Edwin Hubble in 1925 announced his sensational observations of Cepheid variables in the Andromeda Nebula and other nearby nebulae. It had been known for nearly a decade that if a Cepheid could be identified in some faraway object, the distance to that object could be determined from the pulsation period of the Cepheid. According to Hubble, the distance to the Andromeda Nebula was about 930,000 light years, which implied that it could not be part of the Milky Way system. Within a few years, the island universe theory became generally accepted.

Hubble was at the time acquainted with Einstein's cosmological model, which he first referred to in a paper of 1926 in which he suggested a radius of the closed Einstein universe equal to 2.7×10^{10} parsecs or 8.8×10^{10} light years. Hubble did not refer to the galactic redshifts in his paper of 1926, but soon thereafter he turned to the redshift-distance problem which he believed might be solved with data from the powerful 100-inch telescope at the Mount Wilson Observatory. In a now famous paper published

15 March 1929, Hubble argued from an examination of 46 galaxies that their redshifts varied roughly linearly with their distances. On the assumption that the redshifts were caused by the Doppler effect, it meant that $v = c\Delta\lambda/\lambda = Hr$, where the constant of proportionality was approximately 500 km/s/Mpc (1 Mpc = 10^6 parsecs). Hubble did not state the "Hubble law" in this form, and neither did he identify the redshifts as due to a general recession of the galaxies. With the exception of a few observations made by Hubble's collaborator Milton Humason, all the redshift data in the 1929 paper were taken over from Slipher's earlier work.

Although the linear Hubble relation was soon accepted by the majority of astronomers, it was not immediately seen as a revelation. In fact, Hubble's paper received only a handful of citations during the year of 1929 and none of these considered it observational proof of an expanding universe. In a popular account of Hubble's work of June 1929, the esteemed astrophysicist Henry Norris Russell posed the question: "Are the nebulae really flying out in all directions ... so that the universe of nebulae is expanding without limits into the depths of space?" (Russell, 1929). Russell disliked the expansion scenario, which he associated with de Sitter's cosmological model. "It would be premature, however, to adopt de Sitter's theory without reservation," he wrote. "The notion that all the galaxies were originally close together is philosophically rather unsatisfactory." Although aware of the theoretical significance of the redshift-distance relation, Hubble was reluctant to go much beyond observations. In a summary from July 1929, he made it clear that he, much like Russell, preferred a non-recession explanation of the redshifts:

> "It is difficult to believe that the velocities are real; that all matter is actually scattering away from our region of space. It is easier to suppose that the light-waves are lengthened and the lines of the spectra are shifted to the red, as though the objects were receding, by some property of space or by forces acting on the light during its long journey to the Earth" (Hubble, 1929).

One year later, the interpretation changed drastically, not because of new data but because of new theories.

3.4.3. FRIEDMAN–LEMAÎTRE EQUATIONS

In a paper published in 1922, Alexander Friedman, a Russian theoretical physicist barely known outside his country, demonstrated that if the radius of space curvature R was allowed to vary in time, there would be many more solutions to the cosmological field equations than just the static world models derived by Einstein and de Sitter in 1917. For closed models, he showed that the tensor field equations can be boiled down to a simple pair of differential equations for $R(t)$ involving the time derivatives $\dot{R} = dR/dt$ and $\ddot{R} = d^2R/dt^2$. Friedman stated his equations as

$$\left(\frac{\dot{R}}{R}\right)^2 + 2\frac{R\ddot{R}}{R^2} + \frac{c}{R^2} - \Lambda = 0,$$

and

$$3\left(\frac{\dot{R}}{R}\right)^2 + 3\frac{c^2}{R^2} - \Lambda = \kappa c^2 \rho.$$

Whereas in the static models of Einstein and de Sitter, the cosmological constant Λ was linked to the curvature of space, in Friedman's work the constant was a free parameter to be determined observationally.

The momentous importance of Friedman's work was his recognition of possible dynamical world models, including continually expanding models and cyclic models changing between expansion and contraction. For the first class of models, he even anticipated what much later would be called the big-bang universe: "Since the radius of curvature may not be smaller than zero, it must decrease with decreasing time, t, from R_0 to the value zero at time t'. ... The time that has elapsed from the moment when $R = x_0'$ to the moment that corresponds to $R = R_0$, we again call the time since the creation of the world" (Friedman, 1922). Whereas Friedman's analysis of 1922 was restricted to closed spaces, in a companion paper two years later he went further beyond the framework of Einstein and de Sitter by discussing also static and dynamical models of a constant negative curvature. The so-called hyperbolic space was originally introduced by Friedman's compatriot Nikolai Lobachevsky nearly two hundred years earlier (Section 3.1.1).

Friedman's innovative introduction of dynamical world models was primarily a mathematical investigation aimed to chart the homogeneous cosmological models compatible with the field equations of general relativity. On the other hand, he showed very little interest in physical and astronomical data and consequently avoided to state any preference for a particular world model except that he found cyclic models particularly fascinating. Friedman was aware of Slipher's galactic redshifts, which he knew from Eddington's *Space, Time and Gravitation*, but without referring to the redshift problem in his papers from 1922 and 1924. Nor did his papers contain physical terms such as "nebula," "energy," or "radiation." The same was the case in a semi-popular book he published in Russian in 1923 and in which he systematically discussed expanding, contracting, and cyclic world models. This book, *The World as Space and Time*, remained for a long time unknown outside the Soviet Union and was only translated into English in 2014.

The Belgian physicist and Catholic priest Georges Lemaître followed the recent developments in astronomy and astrophysics, including the debate of a possible redshift-distance relation. In a paper in early 1927, written in French, he unknowingly arrived at the same equations for the time-dependent scale factor $R(t)$ which Friedman had communicated five years earlier. Somewhat strangely, he was unaware of Friedman's paper until later in 1927, when Einstein directed his attention to it. From a formal point of view, the only difference was that Lemaître included a term relating to the radiation pressure in the first of the differential equations. However, in striking contrast to Friedman he focused on the expanding solution, which seemed to agree with astrophysical data and hence corresponded to the one and only real universe. Lemaître was the first to realize that the redshifts are a cosmological effect due to the expansion of space. As he explained, if a galaxy emits light when the radius of the universe is R_1 and the light is received when the radius has increased to R_2, the "apparent Doppler effect" is

$$z = \frac{\Delta\lambda}{\lambda} = \frac{R_2}{R_1} - 1.$$

According to Lemaître, it followed that the observed recession velocity was approximately proportional to the distance. For the constant of proportionality, what would soon be known as the Hubble constant, he estimated $v/r = 625\,\text{km/s/Mpc}$, not far from Hubble's value. Although Lemaître's universe of 1927 was continually expanding, it started asymptotically from the closed Einstein universe, which he estimated to have a radius of about 270 Mpc. For this reason, his original model of the universe did not have a finite age and was not of the big-bang type.

The brilliant works of Friedman and Lemaître made no immediate impact on the small cosmological research community dominated by the paradigm of a static universe. Although Friedman's papers were published in the prestigious and widely read *Zeitschrift für Physik*, they attracted almost no attention. Einstein was one of the very few exceptions, and he resolutely rejected Friedman's dynamical models as physically inadmissible if mathematically possible. As regards Lemaître's paper of 1927, it was published in the relatively obscure *Annales Scientifiques Bruxelles* and for this reason alone little known. In fact, it seems to have received no citations at all from other scientists until the spring of 1930. Again, Einstein knew about Lemaître's model, but he dismissed it in the same way that he had dismissed Friedman's paper. Neither of the two innovators of cosmology swayed Einstein or others to seriously consider a universe changing in time. In an article for the 1929 edition of *Encyclopedia Britannica*, Einstein confirmed his position: "Nothing certain is known of what the properties of the space-time continuum may be as a whole. Through the general theory of relativity, however, the view that the continuum is infinite in its time-like extent but finite in the space-like extent has gained in probability" (Einstein, 1929).

Lemaître had sent a copy of his paper to Eddington, his former professor, but the famous astronomer had forgotten about it without reading it. When Lemaître in early 1930 reminded Eddington about the paper, he and other prominent astronomers suddenly realized that the dilemma between the two static world models of Einstein and de Sitter was illusory and that the expanding model provided an answer to the problem. With the enthusiastic support of Eddington and de Sitter, Lemaître's expansion theory moved from obscurity to prominence. According to de Sitter:

> "Lemaître's theory ... gives a complete solution to the difficulties it was intended to solve, a solution of such simplicity as to make it appear self-evident, once it is known ... There can be not the slightest doubt that Lemaître's theory is essentially true, and must be accepted as a very real and important step towards a better understanding of nature" (De Sitter, 1931a).

It took a little longer for Einstein to accept the new paradigm, but by early 1931 he too had converted to the expanding universe. In short, latest by the mid-1930s a majority of astronomers accepted the theory of the expanding universe as a foundation of cosmological research. Whereas Lemaître became celebrated as a pioneer of cosmology, Friedman could not enjoy the late acceptance of his theory. He died prematurely in 1925.

3.4.4. WHO DISCOVERED THE EXPANDING UNIVERSE?

It is generally acknowledged that modern cosmology is crucially founded on the recognition that the universe is expanding, an insight that counts as one of the greatest ever discoveries in cosmology. To whom should this marvelous discovery be credited? The standard answer is undoubtedly that the expanding universe was discovered by Hubble in 1929, but there are also other candidates such as the lesser known Lemaître and Friedman in particular. Although none of the three candidates received the Nobel Prize, two of them were nominated for a physics prize, Hubble in 1953 and Lemaître in 1954. Remarkably, Lemaître was also nominated for the 1956 prize in chemistry, a branch of science he never contributed to and was plainly uninterested in. It took until 1978 before a Nobel Prize was awarded to cosmology, to Arno Penzias and Robert Wilson for their discovery of the cosmic microwave background.

Hubble was nominated by three German scientists in a coordinated but unsuccessful effort. One of the nominators was the eminent German astrophysicist Albrecht Unsöld, who in his nomination to the Nobel Committee argued that Hubble's redshift-distance law of 1929 proved the expansion of the universe. "This work forms the starting point of all later investigations concerning the 'expanding universe' made by Lemaître, Eddington, Einstein, de Sitter, and many others," Unsöld wrote (Kragh, 2017). He further emphasized that Hubble's discovery belonged as much to physics as to astronomy and for this reason was worthy a physics prize: "The discovery that ... in the evolution of the universe a time scale of the order $T = 3$ billion years plays a fundamental role should be assigned an importance similar to the recognition of the fundamental status of the natural constants c and h by Einstein and Planck."

Given that Lemaître's paper dates from 1927 and Hubble's from 1929, Unsöld was obviously wrong in presenting Hubble's work as a starting point for Lemaître. Nonetheless, his contention that Hubble was the true father of the expanding universe has been repeated by later scientists, and even today it appears regularly in textbooks and popular works as were it the authoritative version of history. But it is a poor version, for other reasons because Hubble never claimed himself to have discovered the expanding universe. In 1929, he believed he was conducting a critical test that would allow him to decide between Einstein's and de Sitter's static models. He was not looking for evolving models and made it clear that the linear redshift-distance law was a purely empirical correlation. Sure, he stated the law as a velocity–distance law, thus implicitly interpreting the redshifts as a Doppler effect, but at the same time he emphasized that the involved velocities were "apparent."

Hubble's general attitude was that of a cautious empiricist, not only in 1929 but throughout his career. In his important book *The Realm of the Nebulae* published 1936, he reported that the interpretation of the redshifts as Doppler shifts was generally adopted by theorists and that the velocity–distance relation was now regarded as the observational basis for theories of an expanding universe. However, Hubble thought that this was to go too far, as data did not unambiguously rule out other interpretations. Reluctant to enter what he called the "shadowy realm" of cosmological theory, he wanted to suspend judgment until observations provided an unequivocal answer. Hubble did not deny that the universe expands, but neither was he convinced that it does expand. At some occasions he spoke favorably about the expanding universe, while at others he criticized it as an unnecessary and unproven hypothesis. For example, in an address of late 1941 he said:

> "Red shifts are due either to recession of the nebulae or to some hitherto unrecognized principle operating in interstellar space. The latter interpretation leads to the simple conception of a sensibly infinite homogeneous universe of which the observable region is an insignificant fraction. ... The empirical evidence now available does not favor the interpretation of red shifts as velocity shifts" (Hubble, 1942).

On 8 May 1953, less than half a year before his death, Hubble delivered the George Darwin Lecture in London, and still on this occasion he confirmed his agnostic attitude, avoiding to speak of the expansion of the universe as were it an established fact.

To conclude, if we understand the expansion of the universe in the standard relativistic sense, Hubble cannot be counted its discoverer. This is not only substantiated by Hubble's own statements on the issue but also by the opinions of contemporary authorities none of whom pointed to Hubble as the discoverer of the expanding universe. Neither Eddington, de Sitter, Tolman, nor Milne, nor any other astronomical writer in the 1930s and 1940s, clearly identified Hubble as the scientist who had discovered that the universe expands. Only in the 1950s did the first claims of this kind appear, and by the 1970s they had become the standard story in astronomy textbooks and popular works. Moreover, by then Hubble was receiving the sole credit, whereas the contributions of Friedman and Lemaître, and also of Hubble's collaborator Humason, were ignored or relegated to footnotes. In other words, the shaping of Hubble as the discoverer of the expanding universe is a relatively late historical construction in which Hubble did not himself have any part.

As mentioned, Friedman was the first to demonstrate that expanding models are solutions to the cosmological field equations but without identifying the universe as actually expanding. Although he thus did not discover the expanding universe in the ordinary sense associated with the term discovery, several authors have proposed him as a candidate. The attempts to establish Friedman's priority come mostly, but not solely, from Russian authors who have hinted that the Belgian cosmologist merely followed in the footsteps of the Russian genius. However, there is no documentary basis for the allegation that Lemaître, prior to writing his 1927 paper, knew about and perhaps even plagiarized to some extent Friedman's paper of 1922. In fact, Georges Lemaître is a much better candidate than Hubble and Friedman. As we have seen, Lemaître explicitly predicted the expansion as a relativistic effect in his paper of 1927 with the telling title "Un Univers Homogène de Masse Constante et de Rayon Croissant" (A Homogeneous Universe of Constant Mass and Increasing Radius). On the other hand, he could not justify the derived linear redshift-distance relation with observational data that convincingly supported the relation. Insofar that Lemaître did not establish observationally that the universe is in fact expanding, he did not make a discovery; but insofar as he gave theoretical as well as observational reasons for it, he did discover the expansion of the universe.

It is important to distinguish between the expanding universe and the law of receding galaxies $v = Hr$ or what since the 1950s became known as the Hubble law. As far as the latter law or relation is concerned, Hubble was unquestionably the first to confirm that the galactic redshifts vary proportionally with their distances. His data of 1929 were not quite convincing to all astronomers, but much extended data presented by Hubble and Humason in 1931 left no doubt about the empirical $v = Hr$ law. For the recession constant, they reported $H = 558$ km/s/Mpc or more than eight times the modern value. Although Lemaître had derived a similar law in 1927, he realized that it was poorly justified observationally and he never claimed credit for it. In an article published in 1950, he emphasized that the linear law was Hubble's, whereas he was the first to calculate the recession constant associated with it.

Surprisingly, the issue of the priority to the recession law — but not to the expanding universe — came up during the 2018 meeting of the International Astronomical Union (IAU) held in Vienna. On the initiative of the IAU Executive Council, it was proposed that the long established name Hubble law should be changed to

"Hubble–Lemaître law." The rationale behind the resolution was not very clear, except that the IAU wanted to honor Lemaître for his work. Nor was it clear why the proposed order of names was Hubble–Lemaître and not the chronologically more reasonable Lemaître–Hubble.

Unfortunately, the historical considerations appended to the resolution and in part motivating it were questionable and on occasions at odds with known historical facts. For example, the resolution suggested between the lines that Lemaître had a share in Hubble's work of 1929 because the two met at a IAU meeting in Leiden in the summer of 1928 and supposedly exchanged views about the structure of the universe. Indeed, not only were Hubble and Lemaître both in Leiden during the meeting, so was de Sitter. It may seem natural that Lemaître used the opportunity to tell the two distinguished astronomers about his new paper on the expanding universe, and yet there is not a shred of evidence that he actually met with them. Until such evidence turn up, the supposed conversation with Hubble is pure speculation.

Whatever the objectionable historical background for the 2018 IAU resolution, after an electronic vote among all IAU members, it was passed by a substantial majority. So now astronomers should get used to speak of the Hubble–Lemaître law instead of the Hubble law, at least if they choose to follow the recommendation. So far there is no indication that they do.

3.4.5. TIRED-LIGHT AND RELATED THEORIES

But does the universe expand? Although many astronomers by the mid-1930s would answer affirmatively, not all did. After all, the sole observational basis for believing in an expanding universe was the galactic redshifts and the Hubble law relating them to distances. If the redshifts could be explained by some mechanism different from the one of receding galaxies, it might be possible to retain the static universe, which more than a few astronomers cherished as more natural and intelligible. Throughout the 1930s and continuing after World War II, a variety of such hypotheses were proposed, many of them belonging to the class of "tired light" hypotheses. The name may have been coined in 1932 by the American relativity expert Howard P. Robertson, who was opposed to the hypothesis. It refers to the idea that photons are assumed to lose part of their energy ($E = h\nu = hc/\lambda$) on their long journey through space and therefore arrive at the observer with an increased wavelength. Ideas of this kind appeared shortly after Hubble's 1929 paper and before Lemaître's expanding-universe theory was generally known.

The earliest and most important of the tired-light hypotheses was proposed by the Swiss–American astrophysicist Fritz Zwicky, who is probably best known for his pioneering works on dark matter, supernovae, neutron stars, and gravitational lenses, all dating from the 1930s. Without ever using the term "tired light," in a paper of 1929 Zwicky discussed non-Doppler explanations of the galactic redshifts, focusing on what he described as a gravitational analog of the Compton effect. He calculated that on the assumption of such an effect a photon traveling a distance r from a stationary source would be redshifted by the amount

$$\frac{\Delta\nu}{\nu} = \frac{1.4\,G\rho D}{c^2} r.$$

In this expression, $D >> r$ is a measure of the distance over which the "gravitational drag" operates, and ρ is the average density of matter in the universe. With frequency shifts estimated to be roughly of the same order as reported by Hubble, Zwicky suggested that his hypothesis might be an alternative to the Doppler or velocity interpretation, but he did not yet present it as an alternative to the expanding universe, which was still in the future. In follow-up papers, he discussed observations that might confirm his theory and distinguish it from the rival theory of receding galaxies. Whereas the latter theory was valid for galaxies and clusters of galaxies only, and independent of intergalactic matter, according to Zwicky, the redshifts should turn up also within the Milky Way and depend on the direction of observation. Moreover, it followed from the theory that an initially parallel beam of light would gradually open itself because of small angle scattering. Attempts to detect the Zwicky effect for stars were at first inconclusive and later disconfirming.

Instead of claiming that his own theory was superior to the recession theory, Zwicky admitted that both theories were unsatisfactory, although he thought that his own view was more economical and therefore preferable from a methodological point of view. He recommended cautiousness, not unlike what Hubble did. Astronomers, he said, should not "interpret too dogmatically the observed redshifts as caused by an actual expansion ... but wait for more experimental facts which ... is badly needed before we can hope to arrive at a satisfactory theory" (Zwicky, 1935). The words could as well have been Hubble's. Zwicky's alternative was widely discussed and taken seriously by some leading astronomers, Jeans among them. In the early 1930s, Jeans still hesitated in accepting the expanding universe and for a while he found Zwicky's explanation of the redshifts to be more attractive than the one of Lemaître and Eddington. Only in the late 1930s did Zwicky explicitly reject the expanding universe, which to his mind contradicted observed features of the large-scale distribution of matter.

Although Zwicky's gravitational drag hypothesis was the best known and most elaborate alternative to the relativistic interpretation of the galactic redshifts, it was far from the only one. During the 1930s, more than twenty scientists or amateur scientists suggested alternatives to the expanding universe, many of them belonging to the tired-light category. For example, during the 1930s the famous physical chemist Walther Nernst turned increasingly towards astrophysics and cosmology. Convinced that the universe is static and eternal he suggested a tired-light hypothesis leading to a redshift-distance formula of the form

$$\frac{\Delta \nu}{\nu} = \frac{H}{c} r.$$

While Zwicky's hypothesis assumed photons to interact with intergalactic matter, according to Nernst no such matter was needed. The constant H appearing in his formula was not a constant of the universe, but a "quantum constant" giving the decay rate of photons, $dE/dt = -HE$ or $d\nu/dt = -H\nu$. Other suggestions of a linear redshift-distance relation for a static universe did not rely on tired-light assumptions but on hypotheses of some of the constants of nature, such a Planck's constant h or the speed of light c, varying with time (see Section 1.4.5). Thus, the American physicist Samuel Sambursky suggested in 1937 to reconcile the static universe with the observed redshifts by assuming Planck's constant to decrease slowly in time. To identify the Hubble factor with the relative decrease of h ($H = -\dot{h}/h$), Sambursky was led to the rate of decrease $dh/dt = 10^{-50}$ J.

Most mainstream cosmologists dismissed the alternatives to expansion as speculations based on arbitrary assumptions with no support in known physics. Tellingly, none of the alternatives were published in the leading journals of astrophysics, which at the time were *Astrophysical Journal* (U.S.), *Zeitschrift für Astrophysik* (Germany), and *Monthly Notices of the Royal Astronomical Society* (U.K.). After World War II, even advocates of the steady-state theory, the chief rival to relativistic evolution cosmology in the 1950s, accepted the expanding universe, which was crucial to the steady-state view of the cosmos. Paul Couderc, a French astronomer at the Paris Observatory, expressed the dismissive attitude in unusually strong language:

> "The vanity and sterility of twenty years' opposition to recession is characteristic of a poor intellectual discipline. To hunt for an *ad hoc* interpretation, to search for a means of side-stepping a phenomenon which is strongly indicated by observation simply because it leads to "excessive" conclusions is surely contrary to scientific method worthy of the name. As long as there is no precise, concrete phenomenon capable of casting doubts on the reality of the recession and of explaining the shifts differently, I maintain that it is *a priori* unreasonable to reject recession" (Couderc, 1952, p. 97).

By and large, this has also been the verdict of most later astrophysicists and cosmologists. Tired-light and other non-recession alternatives to the expanding universe are no longer found in high-ranking academic journals devoted to astronomy and cosmology but are relegated to less reputable journals and internet sites. Still, in 1986, the prestigious *Astrophysical Journal* included a paper titled "Is the Universe Really Expanding?," which argued a tired-light alternative to the expanding universe. This may have been the last time a paper of its kind passed the peer-review process.

Part 4

The Universe at Large

4.1

The Age of the Universe

4.1

The Age of the Universe

Even schoolchildren know that the universe was born a finite time ago in a big bang, although they may not know that it happened some 14 billion years ago. We often speak of the age of the universe as were it an age of the same kind as, for example, the age of a bottle of wine or of the age of the Earth. But it is entirely different and, in a sense, most unnatural. For more than two millennia, the consensus view was that things in the universe might have an age but not the universe itself. And if the universe could be assigned an age, it was one based on religious tradition, which offered a value of just a few thousand years. Speculations apart, it was only with the development of dynamical solutions to Einstein's cosmological field equations in the 1920s that the concept of an age of the universe became well-defined and open to scientific investigation. The psychological resistance to the concept is illustrated by the problems some of the greatest scientists of the period, including Einstein and Eddington, had with accepting it.

The earliest big-bang theories, due to Alexander Friedman and Georges Lemaître, included a finite past of a length that could, in principle, be determined, but for a long period of time they were ignored or considered with suspicion. A major reason was the short time-scale of most big-bang models, which made the age of the universe shorter than even the age of the Earth. The time-scale difficulty or age paradox continued to haunt cosmology for three decades and contributed to cosmology's low scientific reputation. The steady-state theory of the universe, which during the 1950s emerged as an apparently viable alternative to the class of relativistic evolution theories, assumed an eternal universe and thus avoided the problem. But there was a prize for it, and by the mid-1960s steady-state cosmology was refuted by astronomical observations. One thing was the recognition that the universe is of finite age, and another was the time that had elapsed since the big bang. While the best answer at about 1950 was less than 2 billion years, today cosmologists agree that the universe was born 13.8 billion years ago.

4.1.1. A WORLD OF FINITE AGE?

According to Aristotle, the world as a whole was necessarily eternal. It had no beginning in time and would never come to an end. Although the main features of Aristotle's philosophical cosmology were generally accepted during Greek-Roman antiquity, there were also critical voices. The atomist school opted for a spatially infinite but temporally finite universe, such as suggested by the Roman poet Lucretius in his famous text *De Rerum Natura* (On the Nature of Things) from about 50 BC. Not only did Lucretius argue that the universe was of finite age, he also thought that it was in a state of decay and would eventually come to a halt. He was of the opinion that "the whole of the world is of comparatively modern date, and recent in its origin, and had its beginning but a short time ago" (Lucretius, 1904, p. 96). It was only with the dominance of Christianity in the early Middle Ages that the belief in a divinely created world of a definite age became established as a dogma. Theophilus of Antioch concluded about 180 AD that creation had taken place in 5529 BC, an approximate age of the universe that would survive until the mid-eighteenth century. By the early seventeenth century, it was generally accepted that God had created the universe at about 4000 BC and also that the Earth had the same age as the stellar universe.

To jump ahead in time, in 1778 the French naturalist Comte du Buffon published a remarkable book, *Époques de la Nature* (The Periods of Nature), in which he reported experiments and calculations aimed to determine the age of the Earth. He arrived at the then staggering figure of about 75000 years, a value which at the time was controversial because it contradicted the still authoritative biblical time scale of approximately 5800 years. Although Buffon restricted his age determination to the Earth, it had cosmological implications given that the universe cannot possibly be younger than the Earth. Thus, if Theophilus were the first to speculate about an exact age of the universe, Buffon were the first to suggest a scientifically based lower limit for its age.

Although ideas of an origin of the universe were generally shunned by scientists in the nineteenth century, there were a few exceptions. The German astronomer Johann Mädler suggested in 1858 to resolve Olbers' paradox (Section 3.1.2) by assuming that the stellar universe had only existed for a limited period of time. However, his suggestion failed to attract serious scientific attention, and the same was the case with the entropic creation argument discussed in Section 1.1.4. Although the line of reasoning based on the second law of thermodynamics indicated a universe with a beginning in time, it was useless when it came to estimating the time at which the entropy of the world began to increase from its minimum state. It should be emphasized that when scientists in the pre-relativity era imagined a beginning of the universe, the "beginning" was a chaotic pre-existing universe consisting of matter and space, but without time or organization. Contrary to later ideas, it did not mean the creation of space and time out of some kind of singularity or super-dense primordial particle.

Until about 1930, the large majority of scientists continued to think of the universe as eternal, if they thought about the question at all. The British geophysicist Arthur Holmes, a pioneer in using radioactive methods in geochronology, estimated in his book *The Age of the Earth* from 1913 that the oldest rocks had an age of 1.7 billion years. On the other hand, he dismissed the entropic creation argument and confirmed that the age of the universe, contrary to the age of the Earth, was infinite. That the notion of a universe with a definite beginning in time was hard to accept, indeed nearly unthinkable, was also illustrated four years later, when Einstein suggested his closed cosmological model based on the general theory of relativity. Not only did Einstein assume a temporally infinite universe, so did de Sitter and other early contributors to cosmology. The assumption was just taken for granted, without discussing it or spelling it out specifically.

In 1924, Herbert Dingle, a young British astrophysicist, innovatively suggested that the redshifts of the spiral nebulae might indicate what he called "the legacy of a huge disruption, in the childhood of matter, of a single parent mass" (Dingle, 1924, p. 399). On this interpretation, it was tempting to conceive "a Universe which had a beginning in time." Although this may look like an anticipation of the big-bang universe, it was not. Dingle's qualitative scenario of a hypothetical past was unrelated to relativistic cosmology, and in the end he concluded in agreement with his peers that there was no evidence for "a beginning of time." He disregarded the argument held by a few scientists, namely that the existence of radioactive substances provided evidence that the world could not have existed for an eternity of time.

In Friedman's earlier but at the time neglected paper of 1922, we have for the first time a clear and quantitative presentation of big-bang world models based on solutions to Einstein's field equations. As mentioned in Section 3.4.3, Friedman introduced the notion of the age of the universe or what he called "the time since the creation of the world." In a brief discussion of expanding solutions, meaning that the scale factor $R(t)$ increases in time, he derived an expression for the time it takes for the universe to expand from the initial state $R = 0$ to its present radius. For a periodic world oscillating between $R = 0$ and a maximum value, he stated that with a zero cosmological constant and a mass of the universe equal to 5×10^{21} solar masses the period would be about 10^{10} years. He did not reveal the source for his estimate of the mass.

Although Friedman realized the hypothetical nature of his models, he could not resist the temptation "to calculate the time elapsed from the moment when the Universe was created starting from a point to its present stage" (Friedman, 2014, p. 80). In his 1923 book *The World as Space and Time,* he suggested that the time was "tens of billions of our ordinary years." Even though Friedman's work was more mathematically than physically oriented, it unquestionably contained the modern concept of the age of the universe.

4.1.2. THE PRIMEVAL-ATOM THEORY

In his groundbreaking paper of 1927, Lemaître assumed a universe expanding as $R \to R_0$ for $t \to -\infty$ and $R \to \infty$ for $t \to \infty$, where R_0 is the radius of the static Einstein world (see Section 3.4.3). There was no proper beginning of the universe, which consequently could not be ascribed a definite age. On the other hand, although Lemaître's model — or what in the 1930s became known as the Lemaître–Eddington model — operated with an infinite past, it did not provide an infinite amount of time for stellar and other evolutionary processes. The Einstein world was a kind of pre-universe out of which the expansion had grown as a result of some instability. In a paper from 1930, Lemaître suggested that the instability had set in some 10^{10} to 10^{11} years ago, a time scale of the same order as earlier mentioned by Friedman. A year later, in a most remarkable note in *Nature* of 9 May 1931 comprising only 457 words, he reduced the initial Einstein world to a much smaller and denser object comparable to a huge atomic nucleus. The title of his note was "The Beginning of the World from the Point of View of Quantum Theory."

Responding to Eddington's view that a sudden beginning of the universe was an unacceptable concept beyond science, such as Eddington expressed it in a popular address of early 1931, Lemaître suggested in his note that "we could conceive the beginning of the universe in the form of a unique quantum, the atomic weight of which is the total mass of the universe ... [and which] would divide in smaller and smaller atoms by a kind of super-radioactive process" (Lemaître, 1931a). Although he spoke of a "quantum" or "atom," what he had in mind was an initial object with a matter density of the order $\rho = 10^{14}\,\mathrm{g/cm^3}$ corresponding to that of an atomic nucleus. At the time, this was the highest imaginable density. Lemaître later changed to another metaphor, describing the primeval atom as an isotope of the neutron. Whatever the metaphor, the hypothetical universe-atom would be subject to the laws of quantum physics and therefore disintegrate, giving birth to space, time, and matter. Because the explosive disintegration was governed by the quantum principle of indeterminacy, it allowed the present world to have evolved from a single, undifferentiated object. As Lemaître phrased it in his note, "The whole matter of the world must have been present at the beginning, but the story it has to tell may be written step by step."

Although Lemaître's hypothesis of a sudden origin of the universe belonged to the class of big-bang models, it did not include the cosmic singularity $R = 0$ at $t = 0$ which was part of Friedman's theory. Lemaître insisted that the starting point was an extended physical object and that the laws of physics would somehow prevent a further gravitational compression of this object. The inspirations for his hypothesis of 1931 are not very clear, but one of them was undoubtedly his reading of Friedman's paper of 1922 and another no less important inspiration came from the existence of radioactive elements. It could be no coincidence, he thought, that the age of the universe was comparable to the lifetimes of uranium and thorium, for had the universe been much older the two elements would no longer exist. He imagined our present world to be the nearly burned out result of a previous highly radioactive universe. In the autumn of 1931, Lemaître presented a more definite and better scientifically argued version of his ideas, now describing his world model as follows:

> "The first stages of the expansion consisted of a rapid expansion determined by the mass of the initial atom, almost equal to the present mass of the universe. ... The initial expansion was able to permit the radius to exceed the value of the equilibrium radius. The expansion thus took place in three phases: a first period of rapid expansion, in which the atom-universe was broken down into atomic stars, a period of slowing-down, followed by a third period of accelerated expansion" (Lemaître, 1931b).

What characterized Lemaître's model apart from the explosive origin was the second phase of slowing-down or "stagnation," a phenomenon which he had first introduced in the context of the Lemaître–Eddington model to explain how the expansion from the Einstein state was caused by condensation processes. The stagnation phase and the theory as a whole depended crucially on the assumption of a positive cosmological constant of which Lemaître was a great advocate (so was Eddington, but for different reasons). The duration of the stagnation phase and hence the age of the universe was given by the quantity $\Lambda_E(1 + \epsilon)$, where Λ_E is the Einstein value

$4\pi G\rho/c^2$ of the cosmological constant. Thus, by adjusting the value of ϵ almost any value of the age of the universe could be obtained. Lemaître often referred to the age as about 10 billion years, but this was just one possibility out of many. Should the universe turn out to be much older, it would not be a problem for the model.

Keenly aware that the primeval-atom theory was hypothetical and might even appear bizarre, Lemaître looked for physical traces that could be interpreted as evidence of the original explosion. He thought that the poorly understood cosmic rays were traces or fossils from the ultimate past. These rays, he speculated, had their origin in the radioactive disintegration of "atomic stars" formed shortly after the initial explosion. However, the hypothesis of cosmic rays as descendants of the primeval atom failed to convince the majority of physicists and astronomers who also rejected or, in most cases ignored, his cosmological theory. As it turned out, a few years before Lemaître's death in 1968, there is indeed a fossil radiation from the big bang, but this is a cold microwave background and not the high-energy charged particles making up the cosmic rays.

While Lemaître's explosion theory received positive attention in the popular press, response from the scientific community was cool and sometimes hostile, tending to ridicule. Eddington never accepted the idea of a beginning of time whether in Friedman's singularity version or in Lemaître's primeval-atom version. While Eddington's opposition was philosophically and perhaps also religiously based, Hubble's reluctance was empirically rooted. In 1936, he compared Lemaître's model with observational data, concluding that although it could not be ruled out it was unlikely as a candidate for the real universe. Hubble found that Lemaître's universe, to agree with observations, needed an unrealistically high density of ca. 10^{-26} g/cm^3 and a radius of only ca. 5×10^8 light years. In an influential review of 1933 on cosmological theories in *Reviews of Modern Physics* H. P. Robertson excluded big-bang models from what he considered plausible models of the universe. He simply ignored the primeval-atom theory.

In 1932, Einstein and de Sitter jointly suggested a theory of the universe which from a formal point of view was a big-bang model if in a quite different sense than Lemaître's. The model was parsimonious in the sense that there was no space curvature, no cosmological constant, and no pressure. It followed from these assumptions and the Friedman equations that the mean density of matter in the universe was

$$\rho = \frac{3H^2}{8\pi G}.$$

With this density, later known as the critical density, the gravitational attraction is precisely balanced by the expansion. Inserting Hubble's value $H = 500\,\mathrm{km/s/Mpc}$, the two scientists obtained the high density $\rho = 4 \times 10^{-28}\,\mathrm{g/cm^3}$. Moreover, the scale factor increases as $R(t) = at^{2/3}$, where a is a constant. Stated in terms of the Hubble time $T = 1/H$, the age of the Einstein–de Sitter universe becomes $t^* = {}^2/_3T$. Remarkably, Einstein and de Sitter did not write down the $R(t)$ variation or the expression for t^*, although these follow directly from the theory. Neither did they note that their model implied an abrupt beginning of the world in $R = 0$ at $t = 0$. These were features that they wanted to pass over in silence. Einstein returned to the model in a little-known article of 1933 in which he analyzed the time-span of the expansion, which he by a numerical error cited as approximately 10 billion years.

The Einstein–de Sitter model was the first well-known open and infinite model of the relativistic universe. It came to be seen as the prototypical big-bang model and played an important role in later cosmological theory. As late as the 1980s, after inflation theory predicted the universe to be flat, it was considered a promising candidate for the evolution of the universe. However, at the time of its publication, neither of the two distinguished authors thought highly about it. In 1932, Einstein reported to Eddington that "I do not think the paper very important myself, but de Sitter was keen on it" (Eddington, 1938). Soon thereafter, de Sitter wrote to Eddington, "You will have seen the paper by Einstein and myself. I do not consider the result of much importance, but Einstein seemed to think that it was" (Eddington, 1938).

4.1.3. A COSMOLOGICAL PARADOX

With the acceptance of the expanding cosmological models, the notion of the age of the universe obtained a firmer but also more problematic foundation. The age could now, for the first time, be related to a measured quantity, the Hubble or Hubble–Lemaître constant, and be calculated for specific models. The result was discomforting because most models resulted in an age of the universe smaller than the age of its constituents in the form of stars, galaxies, and planets. For logical and semantic reasons, the age of the universe *must* be greater than or in principle equal to the oldest of its components, so models not satisfying this fundamental criterion cannot possibly be correct. The problem that haunted cosmology for about three decades is known as the time-scale difficulty or the age paradox. It was a major reason why, until about 1960, expanding models with a beginning in the past were widely regarded with suspicion.

The Hubble time $T = 1/H$ is not the same as the age of the universe t^*, but in most cases it is of the same order of magnitude. Incidentally, since T is a rough measure of the age of the universe, the term Hubble constant is misleading but continues to be used in the scientific literature together with the more appropriate name Hubble parameter. In general, t^* can be computed from T and the time variation of the scale factor, meaning $\dot{R}$ and $\ddot{R}$, but also depends on the average density of matter ρ and the cosmological constant Λ. For $\Lambda = 0$, as was generally assumed, any $\rho > 0$ results in an age smaller than the Hubble time, $t^* < T$. If the density is much smaller than the critical density $3H^2/8\pi G$, the age of the universe approximates the Hubble time. Consider the standard Einstein–de Sitter model for which $t^* = {}^2/_3 T$. The accepted value during the 1930s and 1940s was $H =$ ca. 500 km/s/Mpc or $T =$ ca. 2 billion years, which implies a universe of age $t^* =$ ca. 1.3 billion years. The age of the model universe proposed by Paul Dirac in 1938 was even less, a mere 700 million years or so (Section 1.4.3).

The time-scale difficulty or paradox was that astronomers found stars and galaxies to be much older and that even the Earth was known to be older than 1.3 billion years. As mentioned, as early as 1913 Arthur Holmes concluded that the Earth was at least 1.7 billion years old. Advances in radiometric techniques soon resulted in an even older Earth. Thus, in 1929 Rutherford announced from comparison of the abundances and half-lives of the uranium isotopes U-238 and U-235 that the upper limit of the age of the Earth was 3.4 billion years. After World War II, more refined methods of geochronology resulted in a still older Earth, culminating in the American geochemist Clair Patterson's authoritative value of 4.550 ± 0.070 billion years, which he reported in 1956. Einstein was among those who recognized the scale of the problem. In a review on relativistic cosmology from 1945, he wrote:

> "The age of the universe ... must certainly exceed that of the firm crust of the earth as found from the radioactive minerals. Since determination of age by these minerals is reliable in every respect, the cosmologic theory here presented would be disproved if it were found to contradict any such results. In this case I see no reasonable solution" (Einstein, 1945).

The age paradox appeared in an even more drastic form if the Hubble time were compared to the ages of stars and galaxies. According to James Jeans and the majority of astronomers in the early 1930s, the ages of the oldest stars were as much as 10^{12} years, about one thousand times more than the age of the universe! Although Jeans' long time-scale was soon replaced by a much shorter stellar time-scale of 3–8 billion years, the paradox remained.

Hubble, de Sitter, Robertson, Tolman, and other mainstream cosmologists were deeply worried over the time-scale problem, but without concluding that it forced them to abandon the evolving models based on Einstein's equations. Non-mainstream contributors saw it in a different light and as a reason to support non-recession explanations of the Hubble law, such as the tired-light hypotheses suggested by Zwicky and several other astronomers. To Hubble, it was an additional reason not to embrace expanding models with a definite span of time. To Tolman and also de Sitter, it indicated that the beginning of the expansion should not be automatically identified with the beginning of the universe.

De Sitter suggested in the early 1930s that the evolution of stellar systems might be governed by a time-scale different from that of the universe as a whole. He even speculated that to cut the Gordian knot, cosmologists had to live with the age paradox, in the same way that quantum physicists had learned to live with the paradoxes inherent in quantum mechanics. In his important textbook published in 1934, *Relativity, Thermodynamics and Cosmology*, Tolman despairingly wrote, "Indeed, it is difficult to escape the feeling that the time span for the phenomena of the universe might be most appropriately taken as extending from minus infinity in the past to plus infinity in the future" (Tolman, 1934, p. 486). Nonetheless, although physicists and astronomers recognized the problem, they did not consider it to be insurmountable or something that seriously questioned the credibility of relativistic cosmology. There were various ways in which the problem could be resolved or at least circumvented.

One obvious way, and the natural one if seen in retrospect, would be to revise the Hubble constant on which the calculations of the age of the universe were based. The smaller the Hubble constant, the larger the age. But curiously, until the early 1950s astronomers had such confidence in Hubble's authoritative measurements of H = ca. 500 km/s/Mpc that they rarely questioned them. In about 1950, the generally accepted value was H = 540 km/s/Mpc with an estimated uncertainty of 10%, which corresponded to a Hubble time of 1.8 billion years. Assuming that Hubble's value was correct, cosmologists were forced to look for other ways of dismounting the age paradox. One option was to adopt cosmological models unaffected by the paradox, such as the Lemaître–Eddington nonsingular model or Lemaître's primeval-atom model both of which incorporated the cosmological constant. The latter model could accommodate a universe of age 60 billion years or even more, but very few cosmologists apart from Lemaître himself considered it an acceptable solution to the problem.

In light of the lack of other possibilities, cosmologists could fall back on the flexibility of Einstein's cosmological field equations or the Friedman–Lemaître equations, which provided sufficient room to avoid any glaring inconsistency between theory and data. One possibility, apart from reintroducing the cosmological constant, was to abandon the standard assumption of the so-called cosmological principle, meaning that the universe is homogenous and isotropic. The principle was satisfied by Einstein's original model of 1917 and also by most later cosmological models, but not all. Calculations from the late 1940s demonstrated that for a class of inhomogeneous world models the age came out close to 4 billion years, which Tolman with forced optimism took as evidence that the time-scale problem might not be real after all. Still, it took another decade until the problem disappeared, and even then it did not disappear completely. The story of the age paradox in the period from about 1930 to the mid-1950s illustrates how a paradigm can survive for an extended period of time even when faced with falsifying evidence.

4.1.4. THE STEADY-STATE SOLUTION

In 1953, the *British Journal for the Philosophy of Science* announced a prize competition for the best essay on "the logical and scientific status of the concept of the temporal origin and age of the universe." Twenty-six essays were received and after they had been evaluated by a committee, the first prize was awarded the American philosopher Michael Scriven. Symptomatic of the state and reputation of cosmology at the time, at least in the eyes of philosophers of science, Scriven considered the question of the age of the universe to be unprovable in principle and hence scientifically meaningless: "No verifiable claim can be made either that the universe has a finite age or that it has not. We may still believe that there is a difference between these claims, but the difference is one that is not within the power of science to determine, nor will it ever be" (Scriven, 1954). Another of the essayists argued that the concept of time was inapplicable to the universe as a whole. There is little doubt that the journal's decision of the prize topic was inspired by the new and controversial steady-state theory according to which the universe we observe today has existed in an infinity of time and will continue to exist in another infinity. Indeed, several of the essayists discussed the theory, most of them critically.

The steady-state theory was introduced in two papers of 1948, one written by Fred Hoyle and the other by Hermann Bondi and Thomas Gold. The three young Cambridge physicists agreed that the expansion models derived from the Friedman–Lemaître equations were methodologically objectionable because they could accommodate almost any observation and hence had no real predictive power. Hoyle, in particular, objected to the finite-age expansion models of the Lemaître or Einstein–de Sitter kind, arguing that the hypothesis of a universe created in the past was plainly unscientific. It was in this context that he, on 28 March 1949, coined the name "big bang" in a BBC radio broadcast. The age paradox appearing in the evolutionary models was one of the motivations behind the steady-state theory, if not the only or most important one. Bondi and Gold commented on the paradox as follows:

> "The reciprocal of Hubble's constant, which is a time (*T*), defined by the observations of the velocity-distance law, is between 1800 and 2000 million years. ... Terrestrial observations of radioactive decay in rocks indicate an age of these rocks of at least 2×10^9 years, and some samples seem to imply an age of more than 3×10^9 years. Astrophysical considerations tend to indicate an age of stars in our galaxy of about 5×10^9 years" (Bondi and Gold, 1948).

Bondi and Gold then pointed out that according to standard finite-age models, "the origin of the universe must have been catastrophic and took place less than 2×10^9 years ago. This clearly contradicts the terrestrial and astrophysical evidence." Hoyle's paper included remarks to the same effect. Although the steady-state physicists recognized that the paradox could be avoided by introducing special assumptions, such as large homogeneities or a particular value of the cosmological constant, they considered such saving operations to be illegitimate fine-tuning.

The three physicists wanted an unchanging yet expanding universe, which forced them to the basic assumption that matter, possibly in the form of hydrogen atoms, is continually and spontaneously created throughout the universe if at an imperceptibly low rate. According to the "perfect cosmological principle," there could be no large-scale difference between the past and the future, no cosmological arrow of time. The matter density in the steady-state universe must remain the same, and since the universe expands, matter creation follows as a necessary consequence. This central element in the theory was highly controversial because it seemed to contradict the fundamental law of energy conservation and therefore also the general theory of relativity.

The time-scale problem obviously vanished in the steady-state theory, which was one of its advantages and a main reason why it attracted serious attention. The theory also resulted in several precise and testable predictions and, in this respect, fared better than the rival evolutionary models. Contrary to these models, the Hubble constant was a true constant and the constant Hubble time just a characteristic time-scale that had no connection at all to the age of the universe. Hoyle calculated the average density of matter to be $\rho = 3H^2/8\pi G$, precisely the same as in the critical Einstein–de Sitter universe. Also similar to the Einstein–de Sitter model, the steady-state space was flat or Euclidean and the cosmological constant was zero. As to the rate of matter creation, it came out as $3\rho H$ or ca. 10^{-43} g/s/cm^3. Moreover, the universe advocated by Hoyle, Bondi, and Gold expanded exponentially as

$$R(t) = R_0 e^{Ht} = R_0 e^{t/T}.$$

From the mid-1950s, it became common to characterize the slowing down of the expansion in terms of the dimensionless deceleration parameter q_0, which is defined as

$$q_0 = -\left(\frac{\ddot{R}}{RH^2}\right)_0,$$

where the subscript refers to the present time. Thus, the steady-state model has the value $q_0 = -1$, much lower than the value $q_0 = ½$ of the Einstein–de Sitter model. Contrary to the empty de Sitter universe, which (as understood at the time) also expands exponentially, the steady-state universe was filled with matter at constant density. Yet another steady-state prediction that distinguished it from evolutionary cosmology was that there is no particular era for galaxy formation, but that the average age of galaxies must be equal to $T/3$.

In the heated controversy over the steady-state theory, the ages of galaxies were one of the issues turned against the theory. George Gamow, the chief exponent of big-bang cosmology, rejected the theory of Hoyle and his allies completely. Featuring Hoyle as the author of a cosmic opera, Gamow parodied the steady-state theory in the form of a verse:

> "The universe, by Heaven's decree
> Was never formed in time gone by,
> But is, has been, shall ever be —
> For so say Bondi, Gold and I.
> Stay, O Cosmos, O Cosmos, stay the same!
> We the Steady State proclaim!" (Gamow, 1965, p. 60).

As Gamow was pleased to point out, it follows from the steady-state theory that only 5% of the galaxies have ages larger than T. This means that if most galaxies have an age about or larger than T, the theory will be in trouble. In a paper from 1954, Gamow referred to the recent revision of the Hubble time to about 5 billion years. According to the steady-state theory, this implies that "the nearest galaxy of similar age as the Milky Way should be about twenty times as far from us as the average distance between the galaxies. Thus, since the steady-state theory of the universe does not deny to individual galaxies the right to evolve in time, we should find ourselves surrounded by a bunch of mere youngsters, as the galactic ages go!" (Gamow, 1954a).

The smaller value of the Hubble constant established in the mid-1950s was largely irrelevant with regard to the steady-state theory, in which H or T was just a parameter unconnected to the age of the universe. But indirectly, the revision did affect the theory, not only through a larger value of the mass density (varying as the square of T) but also because it worsened the problem with the galactic ages. At about 1960, the consensus view was that the majority of galaxies had ages of the order of $2T$, which strongly disagreed with the steady-state prediction. This was only one of the observational problems that faced the theory, which at the time was rapidly losing credibility. With the discovery of the cosmic microwave background radiation in 1965 and roughly simultaneous measurements of the abundance of helium in the universe, the steady-state cosmological theory was largely abandoned and relegated to the graveyard of wrong theories. Hoyle refused to admit defeat and continued to his death in 2001 to develop new cosmological models that retained the fundamental idea of an eternal universe. But few cosmologists listened to him. From today's perspective, the steady-state theory belongs to the past and is of more interest to historians and philosophers than to scientists.

4.1.5. FROM 1.2 TO 13.8 BILLION YEARS

During a period of less than seventy years, the universe has aged by more than 12 billion years. It sounds paradoxical but is true enough if the second age refers to astronomers' belief of how old the universe is. To repeat, by 1950 the traditionally accepted recession constant was $H = 540\,\text{km/s/Mpc}$, corresponding to a Hubble time of 1.8 billion years with an estimated uncertainty of about 10%.

The first major step away from this supposedly authoritative value occurred in 1952, at a meeting of the International Astronomical Union in Rome, where the German–American astronomer Walter Baade announced orally that Hubble's time-scale had to be increased to 3.6 billion years or possibly more. He reached the startling conclusion after having critically re-examined Hubble's original use of Cepheid variables to determine the distance to the Andromeda Nebula. One might believe that Baade had then doubled the age of the universe, which was also the conclusion of most astronomers and cosmologists. But Baade thought otherwise, for he — not unlike Hoyle and the steady-state theorists — considered Hubble's T constant to be just a characteristic time-scale for the universe and not a measure of its age. Baade generally disliked cosmology, which he thought was a waste of time and a premature science if a science at all. He consequently was unwilling to relate his discovery to issues of cosmology. Other astronomers were less reluctant, and within a year or two the cat was out of the bag, leading to new measurements that further increased the age of the universe.

After Hubble's death, observational cosmology at the Mount Palomar Observatory was taken over by Allan Sandage, who soon emerged as a leader of the field. Together with Humason and Nicholas Mayall, in 1956 he published a massive paper in which the three authors reported a Hubble constant of $180\,\text{km/s/Mpc}$ with an uncertainty about 20%. The corresponding Hubble time was $T = 5.4$ billion years. Two years later, Sandage proposed a

Hubble constant in the vicinity of 75 km/s/Mpc, which on the assumption of an Einstein–de Sitter universe, where $t^* = {}^2/_3T$, translated into a cosmic age between 6.5 and 13 billion years. Sandage cautiously interpreted the result as support of a big-bang model of the universe: "There is no reason to discard exploding world models on the evidence of inadequate time scale alone, because the possible values of H are within the necessary range" (Sandage, 1958).

The revised time-scale was good news for the big-bang theory and bad news for the steady-state theory. On the other hand, it did not make the age paradox disappear. Although the universe was no longer clearly younger than the Earth, it still appeared to be younger than many of the stars. Theories of stellar evolution from the late 1950s resulted in ages of the oldest stars between 15 and 20 billion years, which thus would reinstate the time-scale difficulty. But were the theoretically derived stellar ages reliable? The blame for the discrepancy between the age estimates of the universe and the stars could be put on the latter, which many astronomers did. Whether for good or bad reasons, at about 1960 the age paradox was declared a non-paradox or at least a potential problem of no great significance. The decision was as much psychological and sociological as scientific, but with ever-increasing observation values of the Hubble time and better estimates of stellar ages the astronomers' intuition proved right. By and large, the time-scale difficulty ceased being an important issue in cosmology.

This is not to say that the further development occurred smoothly or that the controversies over the age of the universe stopped, only that they no longer were considered important to the question of a finite-age universe. During the 1970s and 1980s, specialists disagreed about the value of the Hubble constant and the associated cosmic time-scale. Based on a wide range of data, in 1979 Sandage argued that the Hubble constant was close to 50 km/s/Mpc and the deceleration parameter q_0 close to 0.02, which implied an open universe of age 19 billion years. The long time-scale favored by Sandage was challenged by other experts, notably the French-American astronomer Gérard de Vaucouleurs who maintained that the Hubble constant was about 100 km/s/Mpc and the universe considerable younger than suggested by Sandage. Given that the uncertainties in both cases were about 10%, the values were incompatible.

As it turned out with the emergence of precision cosmology, Sandage and Vaucouleurs were equally correct or perhaps equally incorrect. The Hubble Space Telescope launched in 1990 was designed to measure Cepheid stars as far away as 80 million light years and in this way to narrow down a more precise value of the Hubble constant. The data provided by the satellite resulted in a final value of $H = 72 \pm 8$ km/s/Mpc, but it was only with the reintroduction of the cosmological constant in the late 1990s and the so-called ΛCDM model (Lambda Cold Dark Matter) that precise data for the age of the universe appeared. In the early part of the new millennium, calculations based on the cosmological parameters determined by the successful WMAP satellite — an acronym for Wilkinson Microwave Anisotropy Probe — resulted in an age of 13.72 billion years. Later calculations and observations proved the WMAP result to be robust.

Based on the ΛCDM model, the current best age of the universe is 13.787 ± 0.020 billion years. Given the history of the subject, it is remarkable that today the age of the universe is known as accurately as the age of the Earth. Despite the impressive successes of precision cosmology and astrophysics, the old problem of cosmological versus stellar age has not completely vanished. In 2013, Howard Bond and collaborators reported an age of the star HD 140283 — also and appropriately known as the Methuselah star — of 14.46 ± 0.8 billion years. The Methuselah problem may not yet be finally solved, but no one seriously believes that there can be a genuine mismatch between the two time-scales. As stated in the 2013 paper, "The age of HD 140283 does not conflict with the age of the Universe, given the ±0.8 Gyr [billion years] uncertainty" (Bond *et al.*, 2013). Perhaps there is something wrong with the accepted age of 13.8 billion years of the universe or, more likely, with the age determination of HD 140283.

Although we know that the time from the big bang to the present is approximately 13.8 billion years, not all cosmologists accept that this is the age of the universe as a whole. It is possible, as a minority of physicists speculate, that the big bang was preceded by an earlier universe stretching infinitely back in time just as our universe stretches infinitely forward in time (Section 4.3.5). If so it may be argued that the universe is after all infinite rather than finite in time and that there is no beginning in the absolute sense.

4.2

Questions of Infinity

4.2

Questions of Infinity

When infinities appear in the equations of physics and cosmology, whether in connection with the infinitely small or the infinitely large, it inevitably indicates that something is seriously wrong. Most physicists agree that theories involving infinite quantities are idealizations that cannot represent nature as it really is. The question of infinity not only turns up in connection with the size and age of the universe but also with respect to its beginning in what is formally a singularity, a point at the edge of space-time with an infinite mass density. Such a purely formal oddity first appeared in Friedman's cosmological theory of 1922, whereas a decade later physicists such as Einstein, Lemaître, and Tolman sought to explain it away as nothing but a mathematical construct with no physical meaning. Perhaps, they suggested, could the $R = 0$ singularity at $t = 0$ be prevented if the forces between the tightly packed nuclear particles were taken into account.

The focus of this chapter is on the very large and not on the very small. In a long historical perspective, the idea of an infinite universe is both older and younger than the one of a finite universe, depending on the meaning of the two key concepts "universe" and "infinite." Whereas the standard model of the planetary and stellar universe was for a long time finite, pre-Einstein astronomers generally believed that space was limitless and therefore infinite. However, if space is non-Euclidean, the identity between limitless and infinity is no longer valid and a finite unlimited universe becomes a possibility. For more than a decade, Einstein's closed general-relativity universe, first in a static version and later in an expanding version, was accepted as the best model of the universe at large. Since 1924, when Friedman introduced the possibility of a negatively curved universe, scientists knew, or at least could (and should?) have known, that Einstein's cosmological equations also describe infinite and not only finite world models. Only after World War II were open models widely discussed, in particular in connection with the flat cosmic space predicted by the steady-state theory. Although the steady-state theory was abandoned in the 1960s, flat-space relativistic models became increasingly popular and even more so with the inflation scenarios proposed in the early 1980s. Today, many cosmologists are confident that we live in a flat and presumably infinite universe.

4.2.1. EARLY VIEWS

Whereas the Bible tells us that God created the universe a finite time ago, it has nothing to say about the size of the divinely created universe. In this respect, the views of the ancient Greek philosophers were much clearer if in no way consistent. According to Democritus, the founder of the atomist school, an infinite number of atoms floated incessantly around in empty infinite space. The later Roman author Lucretius agreed, and in his great poem *De Rerum Natura* he argued at length that the idea of a finite universe was absurd. His chief argument was that a finite universe must be bounded, which is impossible unless there is something beyond to limit it; but given that the universe comprises everything, this is nonsensical. About 2000 years later, William Thomson would essentially repeat Lucretius' argument.

On the other hand, the hugely influential Aristotle — and it was his ideas Lucretius criticized — maintained that the universe is finite in matter and space, and so did the no less influential Ptolemy, the great Alexandrian astronomer and mathematician who lived about 150 AD. Presumably because there is nothing in the Bible suggesting an infinite universe, during the Middle Ages and the renaissance era Christian scholars — which at the time meant all scholars — generally adopted Aristotle's view on the size of the universe. For example, spatial finitude was no less part of Copernicus' and Kepler's heliocentric cosmologies than it was of the geocentric cosmology of the ancient Greeks. Not only was Kepler's universe finite, it was unusually small. He concluded that the radius of the sphere of the fixed stars, the outermost part of the universe, was no more than 4 million solar radii and that the volume of the entire stellar sphere was 8 billion times the volume of the Sun. As to the possibility that the universe was infinite, he rejected it on logical and observational grounds.

Despite the consensus view, in the late sixteenth century a few thinkers dared to suggest that there was no limit to the sidereal heavens. To the revolutionary mind of Giordano Bruno, there was no doubt that the physical universe was infinite in size, incorporating an infinity of stars and planetary systems. In a work from 1584 with the characteristic title *De l'Infinito, Universo e Mondi* (On the Infinite, the Universe and the Worlds), he wrote: "There are then innumerable suns, and an infinite number of earths revolve around those suns, just as the seven we can observe revolve around the Sun which is close to us" (Singer, 1950, p. 304). Moreover, Bruno was convinced that intelligent beings populated the planets spread around in the infinite universe. Although the assumption of extraterrestrials was far from new, it was theologically controversial. One of the errors for which Bruno was sentenced to death at the stake in 1600 was his denial that infinity was an attribute possessed only by God.

While the generally accepted view in the seventeenth and eighteenth centuries was that an infinite universe was both philosophically absurd and potentially theologically heretical, there was no consensus with regard to the question. Not a few distinguished scientists and philosophers found the infinite universe to be perfectly acceptable and in better agreement with God's will than a finite one. After about 1685, Newton came to the conclusion that the universally valid law of gravitation required an infinite number of stars in order that the universe would remain mechanically stable. This was what he argued in a famous correspondence of the early 1690s with the learned philologist and theologian Richard Bentley. According to Newton, if the material world were finite and approximately homogeneous on a large scale, all matter would eventually coalesce into one huge central mass, and for this reason he felt forced to adopt an infinite sidereal system.

The infinite Newtonian universe was taken over by Immanuel Kant in his *Allgemeine Naturgeschichte und Theorie des Himmels* (Universal History and Theory of the Universe) from 1755, a work which only became commonly known well into the nineteenth century. It greatly influenced the German astronomer Heinrich Olbers, after whom the famous cosmological paradox of the dark night sky is somewhat unfairly named (Section 3.1.2). Olbers shared Kant's belief in an infinite universe, which he found to be in full agreement with his Christian faith. In his important paper of 1826, he wrote: "Is not space infinite? Is it possible to conceive it to be limited? Can one imagine that the creative power of God should not have made use of infinite space? … In his creative omnipotence the wise God has made the universe slightly opaque, so that we can see only a limited part of the infinite space" (Olbers, 1826). When scientists from Kepler to Olbers spoke of the infinite universe, what they had in mind was essentially a limitless space populated with an incomprehensibly large number of stars.

Philosophical and mathematical questions concerning the infinite are of a general nature and not related to cosmology in particular. Since the time of Aristotle it was common to distinguish between so-called potential and actual (or realized) infinities. Whereas the latter is a determinate totality, such as the complete set of all natural numbers, the first concept is incomplete and closer to indefinite than true infinite; it is a process that approximates the actual infinite. Discussions during the nineteenth century led to the conclusion that if the actual infinite were allowed, it would cause logical paradoxes as well as theological problems. A Catholic, the brilliant French mathematician and theoretical physicist Augustin Cauchy argued that actual infinity was a divine tribute not to be found in nature. "God alone is infinite, outside him everything is finite," he stated in a lecture of 1868 (Cauchy, 1868, p. 27). "Spiritual beings and corporeal beings are given by finite numbers, and the world has its limits in space as well as in time."

As mentioned in Section 1.1.5, the question of the size of the universe was hotly debated in connection with the controversy over the heat death during the last part of the nineteenth century. It was commonly assumed that the second law of thermodynamics only applied to the universe if it were a finite system, which provided a strategy for materialists to dismiss the heat death and its associated argument for entropic creation. By claiming the universe to be infinite, they avoided the unpalatable consequence of a divine creator. While the possibility of an infinite universe was eagerly discussed in relation to the heat death controversy, the inclination of astronomers and most physicists was to declare the question irrelevant and outside the scope of science. Astronomers in the *fin de siècle* period often conceived the universe to consist of a material region, typically identified with the Milky Way system, surrounded by a possibly infinite region of void or ether-filled space. With few exceptions, they were interested only in the region made up of stars and nebulae or other cosmic bodies. Since the space beyond was unobservable, it did not matter much whether it was infinite or just of immense extension.

At that time, there were no good scientific reasons for believing in either a finite or an infinite universe and arguments for one or the other possibility were consequently often of a subjective and rhetorical nature. They reflected what individual scientists intuitively could comprehend or not comprehend. As an example, consider William Thomson's straightforward reason for preferring the infinite over the finite universe: "What would you think of a universe in which you could travel one, ten or a thousand miles, or even to California, and then find it come to an end? Can you suppose an end to matter or an end of space? The idea is incomprehensible" (Thomson, 1891, p. 322). A similar kind of argument could be made with respect to the finite age of the universe, but to Thomson the beginning of everything was not incomprehensible.

The paradox of the finite universe, as Thomson presented it in 1884, arises only if finite space is equated with limited space. This is inevitable if one assumes space to be Euclidean, such as Thomson and almost all contemporary scientists did, but the equation breaks down if space is positively curved. In that case one can travel arbitrarily long distances, even longer than to California, without coming to an end. However, as discussed in Section 3.1.4, the possibility of curved space was practically unknown to astronomers at the time of Thomson's address. When Schwarzschild examined the possibility in his paper of 1900, he was well aware that the finite universe is no less incomprehensible than the infinite one and that a choice can in principle be made on the basis of observations.

4.2.2. HILBERT'S HOTEL

In the early 1880s, the German mathematician Georg Cantor developed a theory of so-called transfinite numbers that provided a new perspective on the problem of the infinite. Cantor concluded that the actual infinite exists in the same abstract sense that finite numbers exist, meaning that the concept is well-defined and operationally useful. Does that imply that the actual infinite can also be part of the physical world? The question was not of great importance to Cantor, who conceived numbers and other mathematical constructs to be no less real than atoms and stars, but it worried other mathematicians closer to the physical sciences. David Hilbert, possibly the most important mathematician of his time, was deeply involved in the creation of general relativity and derived the gravitational field equations, nearly simultaneously with Einstein. Following the developments in theoretical physics closely, after 1925 he immersed himself in the mathematical foundation of quantum mechanics. Although much impressed by Cantor's set theory, Hilbert persistently denied that actual infinities could be found in the real world, which was one of the key messages of a series of lectures that he gave in Göttingen in the mid-1920s.

Hilbert's wide-ranging lectures also covered cosmological issues, which he discussed from the viewpoint of general relativity theory. Not yet aware of Hubble's determination of the huge distance to the Andromeda Nebula, he pictured the stellar universe more or less as Einstein had done in 1917, as a gigantic Milky Way with the myriad of stars filling throughout a positively curved space. In all likelihood, Hilbert was also unaware of Friedman's paper of 1924, in which he had derived hyperbolic and hence spatially infinite solutions from the cosmological field equations. Had he known about the paper, he would presumably have referred to it since it related to his chief concern about real infinities.

Without referring to either the Einstein model or other aspects of cosmology, Hilbert emphasized the crucial difference between finite and infinite sets. As an illustration, in a lecture given on 3 January 1924, he referred to a highly unusual hotel, which since then has been associated with his name.

In a usual hotel with only a finite number of rooms, all of them occupied, there evidently is no way to accommodate new guests. But as Hilbert explained, the situation is strikingly different for an imagined hotel with an infinite number of rooms. Assume that each of the rooms numbered 1, 2, 3 ... is occupied by a single guest. Then:

> "All the manager has to do in order to accommodate a new guest is to make sure that each of the old guests moves to a new room with the number one unit larger. In this way room 1 becomes available for the new guest. One can of course make room for any finite number of new guests in the same manner; and thus, in a world with an infinite number of houses and occupants there will be no homeless" (Hilbert, 2013).

Strange indeed! The hotel can even accommodate a countable infinity of new guests without anyone leaving it. The guests in rooms with the numbers n only have to change to rooms with numbers $2n$, which will leave an infinite number of odd-numbered rooms available for the infinitely many new guests. To Hilbert, the hotel thought experiment was merely an innocent example illustrating that infinities can only turn up in imagined worlds and not in the one we live in. He attached no particular importance to it and nor did it attract contemporary attention. Hilbert's lecture notes referring to the hotel were only published much later. Had the hotel story not been resuscitated by George Gamow more than two decades later, it might have been forgotten.

In 1947, Gamow had just started on his grand project aimed at combining relativistic models of the universe with quantum-mechanical calculations of the formation of primordial atomic nuclei. Within a year, the project would result in the first physical model of the big-bang universe. The same year, Gamow found time to publish a popular book titled *One, Two, Three ... Infinity* in which he covered a broad range of topics ranging from number theory and topology to relativistic space-time and entropy. He also gave his own expanded version of Hilbert's hotel, thereby making it accessible for the first time to a wide readership. Gamow was probably told about Hilbert's lecture when he spent the summer of 1928 in Göttingen, but he did not refer to any source and thus left the readers in doubt of whether the story was authentic or an invention of his own.

Still another version, this time in lyrical form, was offered by an American mathematician in 2006. Imagine a tired driver on a desert highway passing one more hotel with a no-vacancy sign. He nonetheless enters the hotel lobby to see if there should be a room for him:

> The clerk said, "No problem. Here's what can be done —
> We'll move those in a room to the next higher one.
> That will free up the first room and that's where you can stay."
> I tried understanding that as I heard him say:
>
> CHORUS: "Welcome to the HOTEL called INFINITY —
> Where every room is full (every room is full)
> Yet there is room for more.
> Yeah, plenty of room at the HOTEL called INFINITY —
> Move 'em down the floor (move 'em down the floor)
> To make room for more" (Lesser, 2006).

As new guests check in, the driver constantly has to move to rooms with higher numbers, causing him a sleepless night. But to his pleasant surprise, he is compensated economically:

> Last thing I remember at the end of my stay —
> It was time to pay the bill but I had no means to pay.
> The man in 19 smiled, "Your bill is on me.
> 20 pays mine, and so on, so you get yours for free!"

Gamow's book of 1947 also included chapters on stellar nuclear reactions, the age of the universe, and a first presentation of his as yet unpublished nuclear big-bang theory. Convinced that cosmological space was infinite and ever expanding, he implicitly used Hilbert's hotel as an illustration that the ghost of infinity was not menacing to his favored model of the universe. More explicitly, he made the same point five years later, in the semi-popular and widely read *The Creation of the Universe* in which he described how the universe had originated in a hot inferno of nuclear particles. Hilbert's hotel, he said, illustrates that there are no logical problems with a matter-filled expanding or contracting infinite universe. "In exactly the same way that an infinite hotel can accommodate an infinite number of customers without being overcrowded, an infinite space can hold any amount of matter and, whether this matter is packed far tighter than herrings in a barrel or spread as thin as butter on a wartime sandwich, there will always be enough space for it" (Gamow, 1952, p. 36).

Gamow was not the first to point out the possibility of an infinite universe, but at the time he was alone in advocating it as a realistic model in accordance with the equations of general relativity. It is worth noting also that some other models of the universe outside the framework of relativistic cosmology operated with an infinite cosmic space. The most important was the steady-state model, which in almost all respects was the opposite of Gamow's model but shared with it the elements of infinite space and infinitely many galaxies. Arguments relating to but not referring to Hilbert's hotel played some role in the cosmological controversy, if not a very important one. The troubling appearance in the steady-state theory of an actual infinity of objects combined with an infinity of past time caused philosophical concern but were ignored by most physicists and astronomers.

Despite Gamow's attempt to publicize Hilbert's hotel, the paradoxical thought experiment remained for a long time unnoticed. When it turned up at a few occasions, it was in philosophical rather than scientific contexts or it was in connection with infinite time rather than infinite space. After all, the problem of actual infinities relates to any kind of infinity, whether material, spatial, or temporal. Insofar that the hotel metaphor can be used as an argument for infinite past time, or at least that it is a logically possible concept, it should be no surprise that Hilbert's hotel also has turned up in current theological discussions concerning the creation of the universe. By casting doubt on the reality of infinite sets, Hilbert's hotel can be, and has in fact been, used apologetically, as an argument for divine creation.

4.2.3. FROM EINSTEIN TO HOYLE

Kepler's and his contemporaries' estimates of the size of the universe referred to the distance from the central Sun to the sphere of the fixed stars, and the estimates of astronomers in the first decade of the twentieth century were limited to the Milky Way system. To speak in scientific terms of the size of the universe as a whole, including all stars and all space, only became possible after Einstein's revolution of 1917.

As mentioned in Section 3.4.2, in 1926 Hubble argued that the radius of the static Einstein universe was 2.7×10^{10} parsecs or 8.8×10^{10} light years. Other estimates from the 1920s included Knut Lundmark's $R = 3.2 \times 10^{11}$ from 1924 (de Sitter universe) and de Sitter's $R = 1.2 \times 10^{11}$ from 1929 (Einstein universe), both in light years. In a book published in 1930 with the title *The Size of the Universe*, the Polish-Italian physicist Ludwik Silberstein critically reviewed various attempts to determine the global radius of curvature. According to Silberstein, the only model of the universe worth considering was de Sitter's, and for this model he used the Cepheid method to suggest a value of no more than $R = 5 \times 10^9$ light years. Unfortunately, the author still assumed the universe to be static, and so the book was obsolete when it was published shortly after the discovery of the expanding universe.

In the eyes of the public and many scientists, the most important feature of the new cosmology based on general relativity was that it provided a universe of finite size. From about 1920 to 1940, it was generally agreed that the universe, whether static or expanding, must be spherical and hence finite. The belief was not unanimous, though, for cosmology was not yet defined as a branch of general relativity theory. There were in the period several rival views of the cosmos depicting it as infinite and Euclidean, of which Edward A. Milne's theory was the most important. Other alternatives were proposed by William MacMillan in the United States, Walther Nernst in Germany, and Carl Charlier in Sweden, who all disbelieved relativity theory and maintained that the infinite stellar universe was governed by Newton's physics and Euclid's geometry.

Despite flat-space alternatives, the closed universe was generally accepted. It was often considered to be a necessary consequence of general relativity, and that even though it had been known since 1930 that Einstein's theory does not in itself prescribe any particular geometry of cosmic space. To the lay public, the closed Einstein universe might appear emotionally more appealing, perhaps more cozy, than the frightening infinite Newtonian cosmos. This was what the American poet Robert Frost expressed in a poem of 1947:

> "He stretched his arms out to the dark of space;
> And held them absolutely parallel
> In infinite appeal. Then saying 'Hell,'
> He drew them in for warmth and self-embrace.
> He thought that if he could have his space all curved,
> Wrapped in around itself and self-befriended,
> His science needn't get him so unnerved" (Friedman and Donley 1985, p. 78).

Emotional and philosophical associations of this kind were not foreign to the scientists involved in cosmological research. As mentioned in Section 3.1.4, as early as 1900 Schwarzschild preferred a closed universe over an open one, not because observations spoke in favor of the first model but because he found it more agreeable from a philosophical point of view. Lemaître's preference was the same and basically for the same reasons. In agreement with his training in Catholic philosophy, he rejected the actual infinite and remained committed to the finitude of space and matter throughout his career. He strongly believed that the universe, as all its component parts, was comprehensible to the human mind, a belief he found irreconcilable with space populated with an infinite number of objects. In a talk from about 1950 he spoke about "the nightmare of infinite space," declaring that the universe "is like Eden, the garden which had been placed at the disposal of man so that he could cultivate it and explore it" (Godart and Heller, 1978).

The appeal to non-scientific, esthetic or philosophical, considerations such as expressed by Lemaître was quite common at the time, when the choice between different cosmological theories could not yet be decided by observational means. For more than three decades, terms such as "personal taste," "emotional satisfaction," and "philosophical views" appeared abundantly in the scientific literature, not only with respect to finite versus infinite models but also in general. Clearly, cosmology was not yet a science on a par with other sciences, where subjective evaluations had long ago been replaced by objective knowledge based on theoretical and experimental methods.

Infinite space might be a nightmare, but as Friedman first proved in 1924, it was definitely a possibility, namely a solution to the cosmological field equations. Whereas Friedman demonstrated the possibility of open hyperbolic models, he did not include flat or Euclidean models, which were first explicitly mentioned by the German astronomer and cosmologist Otto Heckmann in papers from the early 1930s. As Heckmann pointed out, since there were no observational reasons to believe that the universe was closed rather than open, the choice was left to "philosophical taste." His work inspired Einstein and de Sitter to propose their expanding flat-space model of 1932. Although this model was infinite in space and matter, none of the two authors expressed any concern about the potential problems caused by the appearance of actual infinities in their model universe. The next time a flat-space model appeared on the cosmological scene was in 1938, when Dirac presented his non-relativistic theory based on the Large Numbers Hypothesis (Section 1.4.3). Dirac concluded that the universe could not be closed and since he also ruled out hyperbolic space, he was left with a space of zero curvature. He paid no more attention to the infinities than did Heckmann, Einstein, and de Sitter.

Gamow was yet another of the minority of cosmologists believing in a spatially infinite universe. He first stated the case for a negatively curved space in 1939, in a joint paper on galaxy formation written with Edward Teller, and in his series of works on big-bang cosmology after the war he kept to the hyperbolic universe. Although he realized that there were no compelling reasons to believe in an open universe, he thought that current observational data were best represented by this kind of model. By 1954, he took the relevant data to include a zero cosmological constant, a matter density $\rho = 10^{-30}\,g/cm^3$, and a Hubble constant corresponding to $T = 3.5$ billion years. With these values, "one can calculate that the curvature of our Universe is negative, so that space is open and infinite" (Gamow, 1954b). With the negative space curvature followed an imaginary curvature radius of the universe. Like most earlier cosmologists suggesting an infinite universe, Gamow refrained from commenting on the infinitely many stars and galaxies.

The British astrophysicist Edward Arthur Milne, professor of applied mathematics at the University of Oxford, proposed in the mid-1930s a cosmological model that differed entirely from the models based on general relativity. In sharp contrast to these models, Milne's space was just an infinite Euclidean system of reference devoid of physical properties and therefore could not be curved. Although space could not expand or contract, Milne's universe was expanding in the sense that all galaxies moved apart in accordance with Hubble's recession law. The distance between any two galaxies moving with relative speed v would increase with time as $r = vt$, which implied an age of the universe equal to the Hubble time.

It followed from Milne's unorthodox theory that the universe must necessarily be infinite, meaning an infinite number of galaxies (space itself was neither finite nor infinite). Convinced that the universe was God's creation, not unlike what Olbers and other had done in the past (Section 4.2.1) he associated the infinity of the universe with the infinity of the omnipotent God: "It requires a more powerful God to create an infinite universe than a finite universe. … We rescue the idea of God from the littleness that a pessimistic science has in the past placed upon Him" (Milne, 1948, p. 233). Milne stated that if the universe had been created in a point-singularity, as he believed it was, a system of finite extent was a logical impossibility, something beyond even God's power.

Despite being highly unconventional, Milne's cosmology was taken seriously and attracted great interest, much of it critical. Abstract, deductive, and semi-philosophical as it was, it was of interest also to observational astronomers. Hubble considered Milne's alternative to be valuable and even had some sympathy for it. At a time when Hubble vacillated between a static and an expanding universe, he thought that Milne's system represented both possibilities. However, after World War II, interest sharply declined, and by the mid-1950s the theory was nearly forgotten. What is currently known as the Milne model is a physically unrealistic relativistic model describing an empty universe with negative space curvature. Conceptually, it is completely different from Milne's original model.

The steady-state theory had methodological features in common with Milne's system, and like this it posited an infinitely large universe (Section 4.1.4). Contrary to the Milne theory, it meant a flat space with curvature constant $k = 0$, which followed as a necessary consequence. In other words, could it be proved that cosmic space was curved, the steady-state theory would be wrong. The philosophically troubling appearance of an actual infinity of objects was noted by some of its critics. For example, in 1962 the American physicist Richard Schlegel argued that the theory was contradictory because it led to number of physical objects greater than infinite. A decade later, after the dust from the cosmological controversy had settled, it was suggested that the paradox of Hilbert's hotel was realized in the steady-state universe: "In fact, 'Hilbert's Hotel' describes, metaphorically, the structure of the Universe as it is conceived by the 'Steady State' Cosmology. ... Thus 'Hilbert's Hotel' is no mere mathematical fiction, but, may be the world we actually live in" (Boyce, 1972). However, by then astronomers agreed that we do not actually live in a steady-state universe and so the comparison was of academic interest only.

One can discuss endlessly if the universe is finite or not, but can the question be answered by means of observations? It can, at least in principle. In the mid-1930s, Hubble cautiously concluded from his observations of distant galaxies that while an open universe was ruled out, a closed one was possible if not very likely. That was all. The situation improved in the 1950s when it was understood that the curvature of space can be determined by two measurable quantities, the current Hubble constant H_0 and the deceleration parameter q_0. For models with a zero cosmological constant, the simple relationship was found to be

$$\frac{kc^2}{R^2} = 2H_0^2\left(q_0 - \tfrac{1}{2}\right),$$

which implies open space for $q_0 \leq ½$ (flat for $q_0 = ½$) and closed space for $q_0 > ½$. It was on this basis that Sandage in 1970 wrote a paper with the telling title "Cosmology: A Search for Two Numbers." The idea worked nicely in principle, but data were too uncertain to clearly discriminate between different world models. Although Sandage in 1970 favored $q_0 = 1.2$ corresponding to a closed and finite universe, nine years later his best value had changed drastically to 0.02, an open and infinite universe. Still in the late 1980s, the question was completely open. In the words of the American astronomer Virginia Trimble: "Those of us who are not directly involved in the fray can only suppose that the universe is open ($\Omega < 1$) on Wednesday, Friday and Sunday and closed ($\Omega > 1$) on Thursday, Saturday, and Monday (Tuesday is choir practice)" (Trimble, 1988). The symbol Ω denotes the mass density expressed in terms of the critical density, $\Omega = \rho/\rho_c = 2q_0$ for models with $\Lambda = 0$.

4.2.4. INFLATION AND WHAT FOLLOWED

In the mid-1970s, the best estimates for the mass density of the universe Ω were between 0.2 and 0.6, and later measurements increasingly favored a value not far from $\Omega = 1$. It seemed that we live in an approximately flat universe, perhaps of the Einstein–de Sitter type or something similar to it.

As was realized by the American physicist Allan Guth in about 1980, a roughly flat space today implies that the density must have been incredibly close to the critical value $\Omega = 1$ at the very beginning of the universe. Theoretically, any departure from critical density would increase drastically over cosmic time and, thus, had the initial density been just slightly different, result in a universe very different from the one we observe. Why did the universe begin in this particular critical state? This so-called flatness problem was a major inspiration for Guth's inflation theory of the very early universe, which he introduced in an important paper published in 1981. Other physicists were also involved in the birth of cosmic inflation, in particular the Russian Alexei Starobinsky, who proposed an inflation scenario before Guth.

In a nutshell, according to the theory as proposed by Guth, the universe started in an inflationary phase which, in a split second, blew up — inflated — space by a gigantic factor. Inflation may have started at 10^{-36} s after the initial singularity and ended at about 10^{-32} s. During this incredibly small inflation period of time, space expanded by a factor of 10^{26} or more! Guth's inflation theory provided a mechanism based on particle physics that drove Ω towards 1 and, in this sense, explained the near-flatness of the current universe, something the standard cosmological theory could not do. Another problem that inspired Guth to propose the theory, and which the inflation hypothesis solved, was the apparent absence of magnetic monopoles in the universe (Section 1.3.2). This problem was solved, not by ruling out the hypothetical primordial particles but by making them extremely scarce because of the phenomenal expansion of space during the inflation era.

The original inflation theory or hypothesis was quickly developed into new, more general, and more empirically fruitful versions, which made inflation a widely accepted if still somewhat controversial concept. There are several alternatives to inflation, but so far none of them have succeeded in replacing it as the standard theory of the earliest universe. Between 1981 and 1996, about 3100 papers were published that referred to various aspects of inflation cosmology, and since then the number has further exploded. Today inflation is not just inflation, for there exist a confusingly large number of theories under the inflation umbrella. Indeed, critics have argued that the number of different theories is so great and varied that inflation is practically unfalsifiable.

What matters is that today a majority of cosmologists believe that inflation took place at the birth of the universe and that, as a result of the initial inflationary phase, the universe is flat or very nearly so. According to the current standard cosmological model, the total density parameter Ω is 1 but is made up of different components of which ordinary matter only makes up a little less than 5%. The density is dominated by cold dark matter (about 24%) and dark energy (about 71%) translated into mass by Einstein's $E = Mc^2$ formula. The best measurements tell us that the density parameter is confined to the interval $0.99 < \Omega < 1.01$ and that the initial value must have been almost exactly 1. To be more precise, the density of the very early universe can only have departed from unity by about one part in 10^{62}. Finite cosmological models with $\Omega > 1$ are not ruled out by observations, but the standard view supported by the authority of inflation is that we live in a flat and infinite universe.

If our universe is really of this kind and if it satisfies the cosmological uniformity principle, such as generally assumed, it will contain an actual infinity of protons and also of galaxies. Although there are many more protons than galaxies, both numbers are infinite. (It is hard not to think of infinity as a number, but it is wrong, witness that $\infty + a = \infty$ for any finite number a.) It is tempting to infer, such as some modern physicists have done, that if the universe is truly infinite there must exist an infinite number of copies of everything existing. In the observable finite universe, one proton is a copy of another proton, but one galaxy differs substantially from any other galaxy. According to Max Tegmark, a high-profile Swedish-American cosmologist, "The simplest and most popular cosmological model today predicts that you have a twin in a galaxy about 10 to the 10^{28} meters from here" (Tegmark, 2003). He elaborates:

> In infinite space, even the most unlikely events must take place somewhere. There are infinitely many other inhabited planets, including not just one but infinitely many that have people with the same appearance, name and memories as you, who play out every possible permutation of your life choices. … About 10 to the 10^{92} meters away, there should be a sphere of radius 100 light-years identical to the one centered here.

Alex Vilenkin, one of the pioneers of inflation theory, has similarly stated that it inevitably follows from the theory that every single history, object, or event must be duplicated an infinite number of times. Speculations of this kind, ostensibly based on scientific arguments, are not new. During the controversy over the heat death in the nineteenth century (Section 1.1.5), the French utopian communist Louis Blanqui defended the idea of an infinite cyclic universe in space and time. He asserted that any given moment in time there would be exact copies of any number of humans elsewhere in the universe. These doppelgängers, he wrote in a booklet from 1872, *L'Éternité par les Astres* (Eternity by the Stars), "are of flesh and blood, or in pants and coats, in crinoline and chignon. These are not phantoms: they are the now eternalized" (Blanqui, 1872, p. 47). Far from admitting that he was speculating, Blanqui claimed that his conclusions were "a simple deduction from spectral analysis and Laplace's cosmology."

When cosmologists speak comfortably about the infinite universe, what they have in mind is in most cases just a universe of incomprehensible large dimensions, something quite different from a literally infinite universe. They seem not to be particularly disturbed by the conceptual problems raised by actual infinities. The South African mathematical physicist George Ellis, a former collaborator of Stephen Hawking, is an exception to the relaxed attitude regarding infinities in the real world. In a series of works, he has pointed out that the problems are even more severe today than they were at the time of Hilbert, although they are essentially of the same kind. From Ellis' more philosophical point of view, an actual infinite number of either physical objects or time units in the universe is unacceptable and foreign to testable science. As he writes in a recent paper, "Hilbert's argument would lead us to suppose that the Universe must in fact be spatially finite, with a finite amount of matter in it, a finite number of galaxies and a finite number of living beings" (Ellis *et al.*, 2018).

We know that the observed universe is close to being flat, but we do not know that it is infinite and it is doubtful if we ever will know. Observations can, in principle, tell us that the observed universe is curved ($k = -1$ or $+1$), but not that it has no curvature at all ($k = 0$ precisely). Besides, all such global conclusions rest on the uniformity assumption that the unobservable universe as a whole satisfies the cosmological principle, that is, that it shares the large-scale homogeneity of the universe we know from observations.

.4.3

Oscillating and Bouncing Universes

4.3

Oscillating and Bouncing Universes

The general idea of a universe repeating itself has been considered by philosophers and scientists ever since Greek antiquity and is, even today, taken seriously by a minority of cosmologists. The basic feature of the idea is that the universe oscillates perpetually between creative and destructive phases over immense periods of time. In most but not all models, the oscillations stretch back to the infinite past and forth to the infinite future. Although the different cycles or worlds may differ greatly from one another, they are usually seen as connected, and thus one can still speak of a single universe evolving in endless time. Alternatively, one can speak of a temporal multiverse. While up to the early twentieth century, the cyclic universe was largely a popular philosophical speculation, with the advent of relativistic cosmology the speculation was transformed into scientifically based models of the universe.

There are different names for the idea of a cyclic universe, which is also referred to as the oscillating, periodic, or pulsating universe, names that are usually taken to be synonymous. To qualify as such a universe, the cosmic history has to consist of several full cycles if not necessarily an infinity of them. The simplest cyclic universe is a bang-to-crunch-to-bang universe. A single cycle from a big bang to a big crunch is not enough, although models of this kind are sometimes called cyclic. Moreover, the many and, in most cases, infinitely many cycles cover the past as well as the future cosmic history. It is part of the rationale of the cyclic universe, if not a necessary one, that the present world-cycle has no special significance, since it is just one cycle out of many.

Some modern but unorthodox cosmological models picture the big bang as a transition from a previous universe that has slowly contracted from past eternity. The contraction ends in a big crunch and gives rise to a big bang from which the universe expands into future eternity. Cosmologies of this kind, which have recently been proposed on the basis of string theory and other theories of quantum gravity, are usefully referred to as "bouncing" models. Although they are non-cyclic, they are eternal and share some features with ever-cyclic models, for which reason it is natural to include them under the same umbrella. Yet another name entering in this diverse class of models is the "phoenix universe," which refers to a rebirth of the universe and may be taken to be just another name for the bouncing universe. Somewhat confusingly, the term phoenix universe is sometimes used also for a universe with many rebirths.

4.3.1. PRE-MODERN CYCLIC WORLDS

The notion of some kind of eternally cyclic universe can be found in the earliest mythological world views, whether in Western or Eastern cosmologies. Despite being irreconcilable with Aristotle's conception of an eternal and static universe, it was much discussed in ancient Greek-Roman culture and was particularly popular among philosophers in the Stoic tradition founded by Zeno of Citium around 300 BC. The Stoics conceived the physical universe as a gigantic sphere oscillating through immense cycles of expansion and contraction, or rarefaction and condensation, in the infinite void surrounding it. As to the agent responsible for the eternal changes, they believed it was the element of fire, the action of which would eventually lead to a conflagration or *ekpyrosis* of the entire world. However, this was not the end of the world, for as Chrysippus, a leading Stoic philosopher from around 240 BC, explained, after the catastrophic conflagration the universe would return to its original state. The *ekpyrosis* of the ancient Greeks was a kind of very slow combustion, not a sudden catastrophe. The same idea appeared to the later Roman author Marcus Tullius Cicero, who promoted it in a philosophical treatise called *De Natura Deorum* (On the Nature of the Gods) from 45 BC.

Much later, we find an elaborate argument for the cyclic universe in Kant's cosmological treatise from 1755 in which he presented a universe in perpetual evolution with phases of creation continually followed by phases of destruction (Section 4.2.1). Kant claimed that all celestial structures, from the tiny Earth to the majestic Milky Way, would decay and eventually perish, and yet the inevitable destruction would be counteracted by the creation of new celestial structures elsewhere in the universe. He even speculated that the entire cosmos, or substantial parts of it, might disappear into the original primeval state of chaos and after this catastrophe be born again. The cosmic death-to-life transformation might occur an infinite number of times.

Referring to the old Greek legend of the bird phoenix, Kant wrote about "this phoenix of nature, which burns itself out only to revive from its ashes rejuvenated, across all infinity and spaces … [and] is inexhaustible in new acts" (Kant, 1981, p. 160). Insofar that Kant advocated a cyclic universe, it was eternally cyclic only with respect to the future and not to the past. It would have been theologically dangerous to claim that the cycles continued infinitely back in time. He refrained from speculating about the lengths of the cyclic periods except stating that they were of immense duration. Kant's cosmological scenario remained little known until the mid-nineteenth century, but then its fate changed and it became popular among materialist thinkers in favor of an eternally cyclic universe with no need for divine creation.

During the period from about 1850 to 1910, the cyclic universe was broadly accepted by philosophers and amateur scientists. On the other hand, astronomers and physicists generally eschewed issues of cosmology, and when they did enter the discussion it was often to point out that science offered no support for an eternal universe changing between destruction and regeneration. On the contrary, it was scientifically impossible. As early as 1868, three years after he had introduced the notion of entropy, Clausius concluded that such a view of the universe was ruled out by the irreversible cosmic development prescribed by the second law of thermodynamics (Section 1.1.4). On the other hand, and without mentioning the entropy problem, the astrophysicist Karl Friedrich Zöllner suggested a few years later that in a positively curved cosmic space, all physical processes would occur cyclically (see also Section 3.1.2). Indeed, he thought that the universe itself would be cyclic and that time would go on forever:

> "Parts of a finite quantity of matter moving apart with a finite speed will not reach points infinitely far away. They must converge again after a finite interval of time, the quantity of which depends on the velocity of motion and the curvature of space, and in this way transform, as periodical as a pendulum, kinetic energy to potential energy when they come close and potential to kinetic energy when they separate" (Zöllner, 1872, p. 308).

Zöllner's universe was thus cyclic in the sense that all physical processes occurred periodically, but not in the sense that the size of the universe varied cyclically in time.

The latter possibility was considered by the accomplished Austrian amateur astronomer Rudolf Falb, albeit within the standard framework of Euclidean space. According to Falb, space could not exist without matter, and since he shared Zöllner's belief in only a limited number of stars he was faced with the scenario that in the future all the stars would contract gravitationally and coalesce into what he perceived as a giant sun. However, this would not be the end of the world, for because of the intense heat generated in the contraction process the giant sun would almost instantly evaporate and form an expanding gas. From this gas, new celestial bodies would eventually form. Falb imagined that the contraction-to-expansion process would continue indefinitely and thus secure an eternal universe. He described his cyclic universe as follows: "The life of the world is to be conceived as a recurrence of expansion and contraction, like the breaths of a monstrous colossus. In this way the eternity of processes becomes understandable, that is, the infinite duration of the universe. ... The end of the world is at the same time the beginning of the world" (Falb, 1875).

The concepts of the cyclically recurrent universe discussed by Zöllner, Falb, and their contemporaries differed significantly from the concepts adopted by later generations of cosmologists. What people in the nineteenth century had in mind when referring to the cyclic universe was a world that reprocessed itself by turning apparently dead regions of high entropy into states of low entropy. The universe did not literally disappear to be reborn in the next cycle, such as envisaged in post-1920 cyclic models based on Einstein's theory. Not even Zöllner, who at the time was one of the very few astronomers who knew about and more or less believed in cosmic space being curved, thought about the cyclic world in the geometric sense of later cosmologists.

4.3.2. EARLY RELATIVISTIC MODELS

The cyclic universe reappeared in a radically new form in Friedman's ill-fated paper of 1922 (Section 3.4.3). The Russian physicist not only introduced the expanding model as a solution to the cosmological field equations but also what he called a closed periodic world model with radius increasing from $R = 0$ to a maximum value R_m and then decreasing again to $R = 0$. He recognized that a negative cosmological constant, $\Lambda < 0$, required models of this kind, whatever the value of the curvature. He also recognized that the model was possible if $\Lambda \geq 0$, but then only for positive curvature. Although Friedman's cyclic world of 1922 was just a single cycle from bang to crunch, a year later he compared it to the old speculation of a perpetually oscillating universe, which he found to be particularly fascinating: "The universe contracts into a point (into nothingness), then again, increases its radius from a point to a given magnitude, further again reduces the radius of its curvature, turns into a point and so on. This unwittingly brings to mind the saga of the Hindu mythology about the periods of life" (Friedman, 2014, p. 79). Friedman thus called attention to an eternally cyclic world, but he did not suggest that it was justified either by observations or by the new cosmological equations based on general relativity.

After having belatedly recognized Friedman's work, in the spring of 1931 Einstein proposed his own one-cycle model, which was essentially the same as Friedman's. He first presented the model in a public lecture given at Oxford on 16 May 1931, at a time when his paper had not yet been published. The Einstein or Friedman–Einstein universe follows the simple expression

$$\dot{R}^2 = c^2\left(\frac{R_m}{R} - 1\right),$$

where $\dot{R}$ means the time derivative of the curvature radius. At the middle of the cycle, where $R = R_m$, the rate of expansion is zero. Einstein was pleased that this evolutionary model made the cosmological constant superfluous, but otherwise he did not consider it to be of much interest or to be a good model of the real universe. He was at the time aware of the observations of Hubble (whom he consistently misspelled "Hubbel"), which he used to estimate a mass density of ca. 10^{-26} g/cm^3, a current world radius of ca. 10^8 light years, and an age of ca. 10^9 years. The next year, he replaced the one-cycle model with the non-cyclic and ever-expanding Einstein–de Sitter model.

Einstein's model universe of 1931 was indebted to discussions with the American physicist Richard Tolman, whom Einstein had met during a stay in California and with whom he corresponded. A specialist in cosmology and thermodynamics, Tolman investigated the Einstein model in greater and more physical detail by assuming a universe not only filled with matter but also with radiation. He was the first to introduce thermodynamic considerations in the model, something which neither Einstein nor Friedman had done. Instead of using the classical theory going back to Clausius and Thomson, Tolman made use of a version of thermodynamics modified by the theory of relativity, and from this he derived the surprising result that the expansion and contraction of the model universe were not accompanied by an increase in entropy.

A continual series of successive expansions and contractions — a proper cyclic universe — lacked theoretical justification, but Tolman believed that it was physically possible and even likely. His analysis showed that in this case each new cycle became greater than the previous cycle, both with respect to the period and the maximum value of the curvature radius. The rate of change $\dot{R}$ would also get greater. Moreover, although the entropy would increase from one cycle to the next, it would never attain or approximate a limit of maximum entropy. In other words, the threat of a heat death was canceled. The series of $R = 0$ singularities in Tolman's multicycle model was of course a problem, but he suggested that somehow they were the result of unwarranted simplifications and that they would not appear in a less idealized and more physically realistic theory.

Tolman clearly preferred an everlasting oscillating universe over Lemaître's finite-age big-bang universe, which he viewed with suspicion. On the other hand, he was not philosophically committed to the cyclic universe or, for that matter, any other model of the universe. What really mattered was how well the model agreed with relevant observations, of which there were unfortunately too few to decide between different world models. Tolman fully realized that a mathematical model should not be confused with physical reality, whether in cosmology or in any other branch of science. This was the case for big-bang proposals in particular, but also for his preferred idea of a cyclic universe without an absolute beginning. As he wrote in 1932, "what appears now to be the mathematical possibility for a highly idealized conceptual model, which would expand and contract without ever coming to a final state of rest, must not be mistaken for an assertion as to the properties of the actual universe, concerning which we still know all too little" (Tolman, 1932).

Tolman's penetrating analysis of the cyclic universe helped making it better known and scientifically respectable, but it remained a possibility on the periphery of mainstream cosmology. Tolman may himself have lost confidence in the cyclic universe. From 1934 to his death in 1948, he continued doing work in cosmology, but without paying attention to cyclic models of the kind he had analyzed so carefully in his earlier work.

By the late 1930s, the cyclic universe was well known but faced with several difficulties, observational as well as theoretical. The short time-scale was serious, but it was a problem shared with most other cosmological models. The cyclic model also required a suspiciously high density of matter, although not as high that it was definitely ruled out by observations. On the theoretical side, there was not only the lack of a physical explanation of the bounces from contraction to expansion but also the problem of the number of cycles. To avoid an absolute creation of the universe — and this was one of the reasons why the model attracted attention — there had to be an infinite number of past cycles. It was uncertain if Tolman's multicycle theory could deliver this desideratum. No wonder then that the cyclic universe was not highly regarded. Leading cosmologists including Eddington, de Sitter, and Lemaître either rejected or ignored what they considered to be an unlikely, unnecessary, and speculative hypothesis.

Lemaître considered the Friedman–Einstein–Tolman hypothesis in a paper of 1933, commenting that "solutions where the universe expands and contracts successively while periodically reducing itself to an atomic mass of the dimensions of the solar system, have an indisputable charm and make one think of the Phoenix of legend" (Lemaître, 1933). This may look like support of the cyclic universe, but it was not. Charm, however indisputable, does not indicate truth. Lemaître, the Catholic cosmologist, never supported the idea of an eternal cyclic universe, not even in a version compatible with his own primeval-atom theory.

4.3.3. A FASCINATING BUT MARGINAL IDEA

The cyclic universe attracted renewed attention after World War II, in part for observational reasons, but mostly for philosophical and theoretical reasons. As to observations, the revised Hubble constant of the mid-1950s (Section 4.1.5) implied a longer time-scale and a smaller matter density needed to close the universe. Recall that cosmic space is closed if the density exceed the critical value $3H^2/8\pi G$. Observations reported by Allan Sandage and collaborators in 1956 suggested a closed universe, which Sandage interpreted as potential support for a universe evolving cyclically in time. "If the expansion of the universe is decelerating at the rate the photographic data suggest," he wrote, "the expansion will eventually stop and contraction will begin. If it were to return to a superdense state and explode again, then in the next cycle of oscillation, some 15 billion years hence, we may find ourselves again pursuing our present tasks" (Sandage, 1957). Although Sandage continued to think of such a scenario as a most interesting one, he soon became convinced that it did not, after all, agree with observational data.

Many scientists as well as non-scientists have felt attracted by what they consider the esthetic and emotional appeal of the cyclic universe. One example is Ernst Öpik, an eminent and versatile Estonian-Irish astronomer and astrophysicist who may be best known for his theory of the origin of comets but who also contributed to nuclear astrophysics. Öpik's expressed preference was a kind of temporal multiverse, namely, an eternally cyclic universe in which each new cycle is a new world offering its own variety of events and prospects of evolution. His vision was not the boring eternal recurrence in which each new cycle is an exact replica of the earlier, but an optimistic view of a universe with endless future surprises. From a qualitative point of view, Öpik's oscillating world was not substantially different from what the ancient Stoic philosophers had contemplated. He wrote:

> "Everything will perish in a fiery chaos well before the point of greatest compression is reached. All bodies and all atoms will dissolve into nuclear fluid of the primeval atom — which in this case is not truly primeval — and a new expanding world will surge from it, like Phoenix out of the ashes, rejuvenated and full of creative vigour. No traces of the previous cycle will remain in the new world, which, free of traditions, will follow its course in producing galaxies, stars, living and thinking beings, guided only by its own laws" (Öpik, 1960, p. 123).

Less poetically but more scientifically, the problems of the cyclic universe were investigated in detail by Herman Zanstra, a distinguished Dutch astronomer, who reexamined and extended Tolman's earlier work. Whereas Zanstra confirmed that new cycles would grow increasingly bigger, he also concluded from entropy-based arguments that the oscillating universe could not have existed prior to a certain time. That is, our present world could only have been preceded by a finite number of earlier worlds or cycles. The conclusion was not an argument against the cyclic universe *per se*, but it posed a serious problem for eternally cyclic models. After all, if there were a first cycle, the question of an inexplicable absolute origin reappeared in a version that only differed technically from the one associated with the big-bang theories of Lemaître and Gamow. Zanstra's paper from 1957 was published in the proceedings of the Dutch Academy of Science, and perhaps for this reason it was little known. The results communicated in the paper were independently derived by later physicists and cosmologists unaware of Zanstra's contribution.

During about 1970, after the big-bang theory had become commonly accepted, many physicists concluded that although multicycle models might offer an infinite future, they were bound to have a finite past. In an important textbook originally published in Russian in 1975, the leading cosmologists Yakov Zeldovich and Igor Novikov argued that the cyclic model deserved attention because it could explain the origin of matter as the result of matter in the previous cycle being transformed to hydrogen at the end of the contraction phase. But they also pointed out that the entropy problem remained and that this alone made cyclic models unattractive:

> "The Universe has lived through only a finite number of cycles in the past and has a finite time of existence because in each cycle the entropy increases by a finite amount ... Thus, with due regard for the growth in the entropy, the oscillating model of the Universe does not allow one to describe a Universe with eternal existence from $t = -\infty$. ... The assumption that cycles are possible thus only pushes aside the difficulty concerning the existence of an initial singular state; it does not remove the difficulty" (Zeldovich and Novikov, 1983, p. 660).

Zeldovich and Novikov belonged to the pioneers of standard big-bang cosmology, and they had no confidence in theories with many bangs.

To return to Zanstra, like other cyclic cosmologists he wanted to avoid the unphysical singular states separating the previous contraction from the subsequent expansion. In his paper in 1957, he worried about the mechanism that could produce a bounce just before the mass density became infinite. He found that the temperature would increase dramatically during the final phase of contraction, where all matter in the universe would turn into a very hot and dense gas of protons and electrons. This gas, he suggested without proving it, would act as a stress or negative pressure preventing the singularity and at the same time reverting the contraction. The model that resulted from Zanstra's work was thus a universe undergoing a series of smooth rather than discontinuous transitions. He thought that such a model or picture was not only scientifically justified but also preferable because it agreed with what he called "philosophical desires."

A picture somewhat similar to Zanstra's was proposed by the British physicist William Bonnor, who considered the eternally cyclic universe to be more satisfactory than both the big-bang theory and the rival steady-state theory. Much like Hoyle and some other steady-state advocates, Bonnor considered the beginning of the universe at $t = 0$ to be plainly unscientific and nothing but a cover for divine creation. But he was also convinced that the universe followed the laws of general relativity, and in this respect he rated the big-bang theory more highly than the steady-state alternative. His preferred candidate for a model of the universe was eternally cyclic, with the big crunches and big bangs replaced by smooth transitions. Bonnor promoted his alternative in a popular book published in 1964 titled *The Mystery of the Expanding Universe*, but few astronomers and physicists found his picture of the universe to be convincing.

The hypothesis of a nuclear fluid or something like it with a negative pressure, such as discussed by Zanstra and Bonnor in the context of the cyclic universe, soon resulted in a proliferation of cosmological models. The pressure density is given by $p = -\rho c^2$, where ρ is the density of matter, and it becomes very large at large values of the space curvature. In a paper of 1965, the Polish physicist Jaroslav Pachner presented a singularity-free cosmological model in which the matter density at maximum contraction was at least of the order of the density of atomic nuclei, meaning $\rho \sim 10^{16}\,\mathrm{g/cm^3}$. Later models in the same tradition went much further, assuming a hypothetical negative pressure given by the incredibly high Planck density of approximately $10^{94}\,\mathrm{g/cm^3}$. The idea of a particular state of nuclear matter with a negative pressure was originally proposed by Lemaître in 1934 (see Section 4.4.3), but it only became part of cosmological theory about two decades later.

Whether consisting of just a single cycle, a limited number of cycles, or an infinity of them, the classical cyclic universe presupposed space to be closed. However, according to widely recognized observations in the 1970s, space was probably open, either flat or hyperbolic, which implied that the expansion of the universe could not be reverted. This was the general verdict at the end of the twentieth century, apparently ruling out the possibility of a cyclic universe. And yet, this was not the end of the story.

It remains to be said that not all proposals of a cyclic cosmology in the second half of the twentieth century belonged to the Friedman–Einstein–Tolman tradition or relied on the concept of curved space changing in time. Some of the proposals were merely motivated by a wish to avoid the perplexing problem of an origin of the universe, which was, for example, the main motivation of the French astrophysicist Alexandre Dauvillier, a professor at the Collège de France (and the same who much earlier had mistakenly thought to have discovered element 72, cp. Section 2.4.2). Dauvillier was convinced that any notion of a finite-age universe was unacceptable and that the big-bang theory was metaphysical rather than scientific. Since he also did not believe in the steady-state theory, in the 1960s he presented his own version of a cyclic universe as an alternative mostly based on classical physics. Dauvillier advocated an infinity of cosmic cycles, but in the older pre-relativity sense of energetic cycles occurring endlessly in an infinite and static stellar universe. His ideas did not involve either a closed space or an expanding universe, and consequently they were disregarded by most mainstream scientists.

4.3.4. THE MORE RECENT SCENE

The classical steady-state theory was based on the assumption that the large-scale features of the universe are homogeneous with respect to both space and time. This so-called perfect cosmological principle, together with the consequence of flat cosmic space, ruled out a universe evolving in cycles. On the other hand, there was neither a beginning nor an end of the steady-state universe, and at least in this respect it agreed with the eternal cyclic universe. Although the original model proposed by Hoyle, Bondi, and Gold was abandoned after the discovery of the cosmic microwave background in 1965, Hoyle continued in a series of works to develop radically revised versions of the steady-state theory in order to escape the difficulties of the old theory and yet avoid a universe with a beginning in time.

Collaborating with his former assistant Jayant Narlikar and the British-American astrophysicist Geoffrey Burbidge, this line of work resulted in the 1990s in a comprehensive cyclic model of the universe, what the three physicists called the quasi steady-state cosmology or QSSC for short. The QSSC universe as expounded in detail in a monograph of 2000, *A Different Approach to Cosmology*, relied crucially on the concept of a hypothetical creation or C-field able to produce new matter without violating energy conservation. The creation of matter caused the universe to expand cyclically in a manner somewhat similar to the earlier models proposed by Zanstra and Bonnor, that is, with smooth transitions from one cycle to the next. Contrary to these models, where space had to be closed, the QSSC space was flat and infinite like in the original steady-state theory.

Although on a very long time-scale, the size of the observable universe increased exponentially as in the classical steady-state theory, in the QSSC universe oscillations on a smaller time-scale were superimposed on this expansion. According to Hoyle and his collaborators, a typical cycle would have a half-period of about 20 billion years and our present epoch be at about 14 billion years after the last minimum, of the same order as the age of the big-bang universe. For this value and the present Hubble parameter $H = 65\,\text{km/s/Mpc}$, they derived a present mass density of $\rho = 1.8 \times 10^{-29}\,\text{g/cm}^3$ (however, as a result of the oscillations H and ρ varied in time).

The QSSC alternative claimed to account for all observational data, not only the cosmic microwave background and its temperature variations but also the recent discovery of the accelerated expansion of the universe. The explanations of these and other observations was however quite different from the ones offered by mainstream big-bang cosmology. For example, the microwave background was not a fossil from the cosmic past and the accelerated expansion not due to dark energy as represented by a cosmological constant. In the eyes of the QSSC theorists, the first phenomenon could be explained as starlight thermalized by tiny dust particles in space and the second in terms of the negative-energy C-field. The QSSC theory was not easily falsifiable as it could be modified to accommodate almost any new observation, which was widely seen as a methodological weakness and added to its cool reception. The theory failed to win support within the community of physicists and astronomers, and although attempts to keep it alive continued for a decade or so, by 2010 the QSSC alternative had practically vanished from the scene of science.

At about the same time that QSSC emerged, a very different cyclic theory was introduced by Paul Steinhardt and Neil Turok, two distinguished theoretical physicists. For the first version of their theory, they picked up the old name *ekpyrosis* used by Stoic philosophers to characterize a universe born in fire. Unlike the QSSC, the new cyclic theory of the universe attracted much attention both within and outside the scientific community, and it is still supported by a minority of physicists. A major motivation for the Steinhardt–Turok theory, as it first appeared in 2002, was dissatisfaction with the widely accepted inflation theory of the early universe (Section 4.2.4). Steinhardt was himself one of the fathers of inflation, but after having worked with the theory for nearly twenty years he reached the conclusion that it was all wrong. Inflation never happened and what looked like it had to be something else.

In an address to the American Philosophical Society of 2004, Steinhardt summarized the essential qualitative features of the new cosmological theory as follows:

> "The new cyclic model ... turns the conventional picture topsy-turvy. Space and time exist forever. The big bang is not the beginning of time. Rather, it is a bridge to a pre-existing contracting era. The universe undergoes and endless sequence of cycles in which It contracts in a big crunch and re-emerges in an expanding big bang, with trillions of years of evolution in between. The temperature and density of the universe do not become infinite at any point in the cycle. ... No high-energy inflation has taken place since the big bang" (Steinhardt, 2004).

Contrary to most earlier models of the oscillating universe, in the new theory the observed universe is flat and infinite, and the cycles are not completely independent. Some of the events that are occurring today will determine the large-scale structure of the universe in the cycle to come. The theory provides the infinitely old universe with one coherent history rather than an infinity of disconnected histories as in Öpik's vision of 1960. The cycles do not grow in size, as in Tolman's old theory, but are approximately of the same size and of the same exceedingly long duration. Moreover, Steinhardt and Turok argued that although the entropy increases from one cycle to the next, the entropy density does not and the new model may therefore allow an infinite number of past cycles.

The Steinhardt–Turok theory leads to certain predictions that differ from those of standard inflationary cosmology and therefore can be used to distinguish between the two theories. In particular, if signatures of primordial gravitational waves are found in the microwave background it will confirm inflation but rule out the new cyclic theory. Although gravitational waves were discovered in 2016, these waves were produced by the collision of two black holes and are not messengers from the big bang. Earlier and much publicized claims of having detected primordial gravitational waves were first interpreted as a success of inflation theory, but it soon turned out that the claims were wrong. So far, it is unknown if the big bang gave rise to gravitational waves.

Steinhardt and other physicists are still developing the new cyclic theory (which is not so new any longer), but it is not generally considered to be a serious alternative to standard inflation cosmology. The same is the case with a theory known as conformal cyclic cosmology (or CCC) proposed by the famous mathematician and theoretical physicist Roger Penrose. According to this theory, the universe goes through an endless series of huge cycles or "aeons," each of which decay in a highly dilute soup of radiation. Out of this lifeless chaos comes the beginning of a new cycle in the form of an expanding universe, and so on. Penrose and his collaborators have recently argued that there exists in the presently observed cosmic microwave background distinct signals that are evidence for the conformal cyclic universe. The signals, they assert, are due to gravitational waves generated by colliding black holes in the previous cycle. However, the existence of these signals is controversial and most specialists deny that they are real. Penrose's conformal cyclic cosmology is generally met with skepticism from other cosmologists, or it is simply ignored.

4.3.5. BEFORE THE BIG BANG

Speculations of what preceded the big bang, if anything, are not new and are not restricted to cyclic models. Another possibility is the bouncing non-cyclic universe that only involves a single big crunch immediately followed by a big bang. The first idea of a bouncing universe in the context of relativistic cosmology may have been de Sitter's response to the age paradox of galaxies being older than the universe (Section 4.1.3). Many years before the appearance of fashionable names such as "big bang" and "big crunch," in 1933 de Sitter suggested that the universe might have contracted during an infinite time and, after passing through a point-like minimum, started to expand.

In an address to the Royal Astronomical Society, de Sitter put forward the hypothesis that "the universe contracted to a point at some definite epoch of time, the galaxies passing simultaneously through this point with the velocity of light" (De Sitter, 1933). He added that this was a mathematical idealization and if the gravitational forces between the galaxies were taken into consideration, "the point becomes a finite region and a physical interpretation is possible." De Sitter speculated that the galaxies survived their hazardous passage through the minimum, a critical event which he dated to approximately 5 billion years ago. According to this scenario, the ages of the galaxies could be much longer than the age of the universe as estimated from Hubble's recession measurements. De Sitter strongly disliked the alternative speculation of a cyclic universe, admitting that his preference for a single bounce was what he called "a purely personal idiosyncrasy."

When de Sitter speculated about a bounce in the early 1930s, big-bang cosmology in the modern sense was still in the future. Apart from Lemaître's early and ill-fated primeval-atom hypothesis, the concept of a hot big bang as the beginning of the universe was essentially due to Gamow and his collaborators in the late 1940s. In one of his early papers on big-bang cosmology from 1948, Gamow imagined how the original state of matter might have been the result of a hypothetical collapse of the universe preceding the present expansion. The extremely high pressure at the collapse might have squeezed free electrons into protons and formed a hot fluid of neutrons, he speculated. Gamow was fascinated by the idea that the universe had evolved from infinite rarefaction to a superdense state and in the far future would again be infinitely rarefied. However, he realized that although the hypothesis of a universe before the big bang was emotionally attractive, it was a dream with no foundation in physics:

> "Mathematically we may say that the observed expansion of the universe is nothing but the bouncing back which resulted from a collapse prior to the zero of time a few billion years ago. Physically, however, there is no sense in speaking about the 'prehistoric state' of the universe, since indeed during the state of maximum compression everything was squeezed into the pulp ... and no information could have been left from the earlier time if there ever was one" (Gamow, 1951).

The kind of pre-big-bang scenario that Gamow judged to be interesting but unscientific was taken seriously by several later physicists, who did not accept Gamow's assumption that "no information could have been left from the earlier time." Since the 1980s, a large number of physicists have proposed theories of how the big bang emerged from a previous big crunch. While some bouncing cosmologies are based on classical relativity with its three space dimensions, others rely on the equations of the many-dimensional theory of superstrings. String cosmology has since the early 1990s been developed by the Italian physicists Gabriele Veneziano and Maurizio Gasperini in particular. In a popular paper published in 2004, Veneziano presented the pre-big-bang string scenario in words strikingly similar to Gamow's speculation more than half a century earlier: "According to the scenario, the pre-big-bang universe was almost a perfect mirror-image of the post-bang one. If the universe is eternal into the future, the contents thinning to a meager gruel, it is also eternal into the past. Infinitely long ago it was nearly empty, filled only with a tenuous, widely dispersed, chaotic gas of radiation and matter" (Veneziano, 2004). In other words, from nothing over something to nothing.

Although the string universe did not literally begin, in the infinite past it was in a cold, flat, and nearly empty state, which itself is unexplained and supposedly in no need of explanation. After all, one has to start somewhere and with something. The initial vacuum-like state developed into a universe with increasing density and curvature until the curvature radius attained a length of the order of the Planck length 10^{-35} m, which according to string theory is the smallest possible length. Our universe emerged in a big bang out of this non-singular and maximally compressed state. In its modern versions, string cosmology has been developed to quantitative models with predictions which can in principle be compared to observations. The relevant observations are primarily details in the cosmic microwave background, where string cosmology predicts distinct traces of gravitational waves. These and other predictions have not been detected, but they are expected to be within the reach of experiments in the near future.

The existence of a smallest length and a corresponding smallest volume of space of the order $10^{-105}\,m^3$ is not only a feature of string theory but also of so-called loop quantum gravity or LQG (Section 1.4.2). Physicists working within the LQG research program have applied it to cosmology and produced elaborate mathematical models of the universe as it would look like if LQG is assumed to be correct. This approach known as loop quantum cosmology or LQC was pioneered by the German-American physicist Martin Bojowald in 1999 and has since been developed by groups of other physicists. LQC results in models that, in some respects, differ from those derived from string theory, but they have in common with these models that they picture the universe as eternal, with no beginning and no end in the absolute sense. According to LQC, the big bang was the result of a non-singular transition from contraction to expansion, either the latest of an endless series of cycles or perhaps just a single bounce in the infinite history of the universe.

Whatever the future will show, there are presently a large number of theories which have in common that although they accept the big bang they deny that it was the beginning of everything. According to these theories — none of which are accepted by the majority of cosmologists, it should be emphasized — the universe as a whole has existed in an eternity, and it will continue to exist in another eternity.

4.4

Plentiful Nothingness

4.4

Plentiful Nothingness

It may appear contradictory to speak of a physical vacuum, for how can a vacuum — a state in which nothing exists — possibly be ascribed physical and, in principle, measurable properties? It can, but only because the physicists' vacuum is entirely different from the nebulous concept of nothingness, which belongs exclusively to the domains of metaphysics and theology. The vacuum of the twentieth century is full of energy and no less real than gross matter. Although this recognition is due to quantum mechanics and general relativity theory, precursors to the idea can be found in the ether theories that dominated fundamental physics in the late nineteenth century. These theories shared with the modern ones that they denied the existence of a completely empty space and instead conceived vacuum to be filled with a medium with a non-removable intrinsic energy. There is some similarity between the classical and long-discarded ether and the presently accepted physical vacuum. As Robert Dicke wrote in a paper on gravitation from 1959: "One suspects that, with empty space having so many properties, all that had been accomplished in destroying the ether was a semantic trick. The ether had been renamed the vacuum" (Dicke, 1959).

The historical process leading to the modern idea of dark energy can be reconstructed as following two very different paths that eventually merged into a single one. One of the paths was rooted in quantum theory and its associated concept of zero-point energy, which first turned up in 1911 and later was taken to imply that even empty space must have a non-zero energy ground state. The other path was cosmological and focused on the physical meaning of the cosmological constant appearing in Einstein's field equations of 1917. As it turned out in the 1930s, the controversial constant was equivalent to a gravitational vacuum energy density associated with a negative pressure. The cosmological constant was, for a long time, unpopular and only seriously returned to cosmology in the 1990s. With the discovery that the universe expands at an increasing rate, Einstein's constant was revived from obscurity to become a most fundamental quantity in the new picture of the universe described by the so-called ΛCDM concordance model. According to this picture, not only does the universe accelerate, it also consists mainly of a dark vacuum energy responsible for the acceleration. Behind it all lies the cosmological constant, which Einstein decided was an unfortunate mistake. Today, we would consider his dismissal of the constant to be a much bigger mistake.

4.4.1. THE UBIQUITOUS ETHER

According to Aristotle, void space or vacuum was about the same as nothingness. By means of thought experiments and logical arguments, he proved to his own satisfaction (and to that of many others) that vacuum could not be part of the natural world, whether on Earth or in the heavens. While the world below the Moon consisted of a plenum of the four elements air, water, earth, and fire, the heavenly world, likewise a plenum, was entirely made up of a fifth element which Aristotle called ether. The vast expanse between the stars and the planets appeared to be empty, but in reality it was filled up with the subtle ethereal medium, what in Latin became known as *quinta essentia*. Contrary to the terrestrial elements, the ether was pure and incorruptible, with no physical attributes except that it was transparent. Since the ether was only found above the Moon, its existence was based entirely on philosophical and astronomical arguments.

Later, philosophers in the Stoic tradition agreed that there could be no empty state within the material world, but instead of filling space with the passive ether they postulated an active *pneuma*, which they conceived as a vital and dynamic principle or substance. While Aristotle's ether was restricted to the heavens, the pneuma also filled up everything in the terrestrial region. The Aristotelian doctrine of *horror vacui* (nature abhors a vacuum) was much discussed by philosophers and theologians in the Middle Ages. Although they agreed that a void cannot exist naturally, they were less certain if God could have created one if he so wished. After all, God is omnipotent and can create anything that is logically possible. Was the vacuum just missing from nature, or was it a self-contradictory concept? Until the scientific revolution in the early seventeenth century, natural philosophers generally accepted the non-existence of a vacuum. Barometer experiments performed by Evangelista Torricelli, Blaise Pascal, and others around 1650 proved that air was a substance with a surprisingly large weight and pressure, and the experiments suggested that a complete evacuation of air would result in a kind of vacuum.

With the sensational invention of the first air-pumps in the 1650s, first by Otto von Guericke in Germany and then by Robert Boyle and Robert Hooke in England, the artificially produced vacuum appeared in a more dramatic and convincing form. Boyle realized that even the most completely evacuated space might not be completely empty, and he refrained from identifying it with the traditional and, to his mind, metaphysical meaning of vacuum. Perhaps, he and others speculated, the void produced in the air-pump might contain some kind of undetectable ethereal substance which was part of space *per se*. The existence of ether or "subtle matter" present in a void soon became the consensus view of natural philosophers. In the 1706 edition of his famous work *Opticks*, Newton speculated at length on the nature of the ether, which he suggested was indispensable to both physics and astronomy:

> "Is not the Heat ... convey'd through the *Vacuum* by the Vibrations of a much subtiler Medium than Air, which after the Air was drawn out remained in the *Vacuum*? And is not this Medium exceedingly more rare and subtile than the Air, and exceedingly

> more elastic and active? And doth it not readily pervade all Bodies? Is not the Medium much rarer within the dense Bodies of the Sun, Stars, Planets and Comets, than in the empty celestial Spaces between them?" (Newton, 1952, p. 34).

While the ether of Newton and subsequent generations was mechanical in nature, with the acceptance of Maxwell's new theory of electromagnetism in the last quarter of the nineteenth century it became inextricably associated with the electromagnetic field. The ether was part and parcel of Maxwell's theory, and Maxwell himself firmly believed in it. In an important paper published in 1864, he speculated that the electromagnetic ether might somehow be responsible for the force of gravity, and in this connection he concluded that the ether must possess "when undisturbed, an enormous intrinsic energy" (Maxwell, 1973, Vol. 1, p. 571). At the turn of the century, few physicists thought of empty space as really empty. It was generally agreed that it was filled with the same electromagnetic ether which explained optical phenomena, intermolecular forces, and the enigmatic electron. Fundamental physics in the *fin de siècle* period was, to a large extent, ether physics. Although pure ether could not be isolated or directly investigated, it existed in the form of empty space. According to one physicist, space without ether was a contradiction in terms, an oxymoron like a forest without trees.

The British physicist Oliver Lodge, a devoted follower of Maxwell and a pioneer in radio telegraphy, was an enthusiastic protagonist of the physically active Victorian ether, which he considered to be incompressible and the reservoir of an immense amount of energy. This energy was not directly detectable, but it could be inferred from the ether's mass density as calculated on the basis of electromagnetic theory. In a calculation dating from 1907, Lodge estimated the minimum density of the ether to be "something like ten-thousand-million times that of platinum," or of the order 10^{12} g/cm^3. A similar value was obtained by J. J. Thomson, the discoverer of the electron. Lodge expressed his conclusion in terms of the energy associated with the ether:

> "The intrinsic constitutional kinetic energy of the æther, which confers upon it its properties and enables it to transmit waves, is thus comparable with 10^{33} ergs per c. c.; or say 100 foot-lbs. per

> atomic volume. This is equivalent to saying that 3×10^{17} kilowatt-hours, or the total output of a million-kilowatt power station for thirty million years, exists permanently, and at present inaccessibly, in every cubic millimetre of space" (Lodge, 1907).

No wonder that Lodge, in an address to the Royal Institution the following year, characterized the intrinsic energy of the ether as "incredibly and portentously great." Although his stated energy density was extreme, he was not the only British ether physicist to support an electromagnetic ether densely packed with energy. For example, in an address of 1896, the Irish physicist George FitzGerald, another expert in Maxwell's electromagnetic theory, estimated a value of 10^{22} erg/cm^3 for the ethereal energy.

To Lodge and some of his contemporaries, the ether was not only of fundamental importance to physics, it was equally important to an understanding of spiritual and religious phenomena. It was a medium intermediary between the material and the spiritual, and Lodge tended to believe that it might be wholly immaterial. A believer in the immortality of souls and an active member of the Society for Psychical Research established in 1882 to investigate paranormal phenomena, he came to see the ether as a link between the physical and non-physical cosmos. As he expressed it in a book of 1925, the ether is "the primary instrument of Mind, the vehicle of Soul, the habitation of Spirit. Truly it may be called the living garment of God" (Lodge, 1925, p. 179). While Zöllner associated psychic phenomena with a fifth space dimension (Section 3.1.3), Lodge considered the ether of space to be the connecting link.

4.4.2. NEVER AT REST

According to quantum mechanics, the energy of a vibrating electron or some other particle can never be zero. Even at absolute zero temperature, the particle will move with a "zero-point energy," which cannot possible be brought to cessation. This remarkable result appeared in the famous paper of September 1925 in which young Werner Heisenberg introduced the new theory of quantum mechanics. Heisenberg found that the energy of a one-dimensional harmonic oscillator with frequency ν was given by

$$E_n = (n + ½)h\nu,$$

where $n = 0, 1, 2, \ldots$ and h is Planck's constant. Thus, even in its lowest energy state, $n = 0$, the oscillator will have the energy $E = ½h\nu$. When Heisenberg announced the result of a zero-point energy, it had been known for more than a decade. The concept was first proposed by Max Planck in 1911 as part of his so-called "second theory," a revision of his original quantum theory of 1900 where the energy levels of the harmonic oscillator were quantized as $E_n = nh\nu$ and thus with $E = 0$ in the ground state. The second theory was important but unsuccessful, and after a few years it was abandoned. Nonetheless, the concept of zero-point energy survived as an interesting if controversial possibility.

Could the zero-point energy be proved experimentally or was it just a theoretical artifact? The question was much discussed by leading physicists, among them Einstein and the Dutch experimentalist Heike Kamerlingh Onnes, a Nobel Prize laureate of 1913. Einstein at first believed that measurements of the specific heat of hydrogen molecules supported the hypothesis of a zero-point energy, but he soon came to the conclusion that the hypothesis was probably wrong. Other physicists looked for evidence in isotope separation or in low-temperature experiments near the absolute zero of temperature. However, the results obtained in Onnes' laboratory in Leiden and elsewhere were ambiguous and not generally accepted as support for the hypothesis. Perhaps the effect existed, perhaps it did not.

The first solid evidence for a zero-point of energy was obtained in early 1925 by the American chemist Robert Mulliken, who in studies of molecular spectroscopy demonstrated that the puzzling zero-point energy really existed (see also Section 1.2.4). By 1926, the hypothesis originally proposed by Planck was no longer controversial insofar as it concerned atoms, molecules, and other material systems. Not only was the hypothesis convincingly confirmed by experiments, it was also required by the authoritative theory of quantum mechanics.

Did zero-point energy also apply to free space or space occupied by radiation alone? Specialists in quantum mechanics said no, but the chemist Walther Nernst said yes. As early as 1916, Nernst proposed in a communication to the German Physical Society that empty space — or ether, as he conceived it to be — was filled with electromagnetic zero-point radiation. He emphasized that even without radiating matter, there would be stored in the ether-filled space an enormous amount of zero-point energy, the density of which he went on to calculate. Nernst argued that the energy density of zero-point radiation at a particular frequency ν was approximately given by $\rho_\nu = a\nu^3$ with $a = 8\pi h/c^3$. Integrated over all frequencies, the result would be infinite, for which reason he introduced a maximum frequency ν_m of 10^{20} Hz corresponding to a minimum wavelength of 3×10^{-10} cm. Integration then gave

$$\rho = \frac{2\pi h}{c^3}\nu_m^4 = 1.52 \times 10^{23}\ \mathrm{erg/cm^3}.$$

Translated to mass density by $E = mc^2$, the result is $150\,\mathrm{g/cm^3}$. Nernst commented that "The amount of zero-point energy in the vacuum is thus quite enormous, causing ... fluctuations in it to exert great actions" (Nernst, 1916). Enormous it was, although not quite as enormous as Lodge's earlier estimate of $10^{12}\,\mathrm{g/cm^3}$, to which Nernst did not refer. Nernst further showed that if zero-point radiation enclosed in a container is compressed, neither its energy density nor its spectral distribution will be affected, a result which he characterized as "truly remarkable."

The ideas that Nernst entertained with regard to the ether and the zero-point energy were known to German physicists without attracting much attention. His proposal of assigning a zero-point energy to pure radiation dissociated from matter was generally dismissed as unrealistic. Indeed, apart from a few references to Nernst's hypothesis of vacuum zero-point radiation in the 1910s and early 1920s, it was effectively forgotten. With the emergence of quantum mechanics and quantum field theory, it fared no better. Specialists in quantum mechanics unanimously agreed that although quantization of the electromagnetic field involved zero-point energies, these were unmeasurable and therefore of no physical significance. What quickly became the consensus view — namely that zero-point energy could be ascribed to material systems only and not to the free electromagnetic field — was expressed by Pascual Jordan and Wolfgang Pauli in a joint paper of 1928:

> "Contrary to the eigen-oscillations in a crystal lattice (where theoretical as well as empirical reasons speak to the presence of a zero-point energy), for the eigen-oscillations of the radiation no physical reality is associated with this "zero-point energy" of $\frac{1}{2}h\nu$ per degree of freedom. We are here dealing with strictly harmonic oscillators, and since this "zero-point energy" can neither be absorbed nor reflected ... it seems to escape any possibility for detection. For this reason, it is probably simpler and more satisfying to assume that for electromagnetic fields this zero-point radiation does not exist at all" (Jordan and Pauli, 1928).

At around 1930, this verdict of two of the specialists in quantum mechanics was generally accepted, but as it turned out, it was not the final answer to the question of the zero-point energy associated with electromagnetic radiation. Ten years later, it had become part of standard physics that zero-point oscillations in empty space are inevitable and can never cease completely. An absolute vacuum in the sense of classical physics is impossible. Or to phrase it differently, the physical vacuum is entirely different from nothingness. On the other hand, there still was no experimental evidence that vacuum zero-point energy and its related fluctuations really existed.

A few physicists in the 1920s contemplated if the issue might be of cosmological relevance, such as Nernst thought it was. In a lecture given in 1912, he appealed to the hidden energy of the ether as a means to save the universe from the heat death, and in the early 1920s he introduced the zero-point energy of the infinite space in his unorthodox speculations concerning a steady-state universe with continual creation and destruction of matter (see also Sections 2.5.1 and 3.4.5). In the context of more conventional cosmology, the German physicist Wilhelm Lenz examined in 1926 the equilibrium between matter and radiation in Einstein's closed and static universe. He found that the unknown temperature of radiation varied with the curvature radius as $T^2 \sim 1/R$ but was unable to obtain a realistic match between T and R. For example, using de Sitter's value R = ca. 10^{26} cm, the temperature of space came out as approximately $T = 300\,\text{K}$.

Referring to Nernst's work, Lenz also considered the role of the vacuum energy, should it exist. However, he concluded that if the space zero-point energy was included, it would act gravitationally by drastically increasing the curvature of space and hence making the universe ridiculously small. He stated his conclusion as follows:

> "If one allows waves of the shortest wavelengths $\lambda \cong 2 \times 10^{-11}$ cm, ... and if this radiation, converted to material density ($u/c^2 \sim 10^6$), contributed to the curvature of the world — one would obtain a vacuum energy density of such a value that the world would not reach even to the Moon" (Lenz, 1926).

To avoid such an absurdity, he decided to ignore Nernst's questionable hypothesis of zero-point energy in empty space. Although Lenz's paper made no impact at the time and was ignored by most cosmologists, it deserves attention because it was the first time that the vacuum zero-point energy appeared in relativistic cosmology.

4.4.3. THE COSMOLOGICAL CONSTANT

The physical meaning of the cosmological constant Λ that Einstein had introduced in his cosmological field equations of 1917 (Chapters 3.4 and 4.1) remained a puzzle for a very long time. Of course, if the constant was a mistake, such as widely believed for about half a century, there was nothing to puzzle about. One reason for the uncertainty concerning the cosmological constant was that next to nothing was known about its magnitude except that it must be very small. In a letter to Einstein of 1917, de Sitter estimated the constant to be smaller than $10^{-45}\,\text{cm}^{-2}$, and probably smaller than $10^{-50}\,\text{cm}^{-2}$. Seventeen years later, in his pioneering textbook *Relativity, Thermodynamics and Cosmology*, Tolman concluded from astronomical observations that the value of the cosmological constant was probably limited to the interval $-2.2 \times 10^{-44} < \Lambda < 2.2 \times 10^{-44}\,\text{cm}^{-2}$. Yet he and other cosmologists realized that it was nothing but an educated guess.

While Einstein abandoned the cosmological constant in 1931, other leading cosmologists including Eddington, de Sitter, and Lemaître found it to be indispensable. De Sitter was convinced that the constant was responsible for the expansion of space. "The expansion depends on the *lambda* alone," he stated in a popular article of 1931, admitting that the mechanism by which the constant operated was unknown (De Sitter, 1931b). In the spirit of positivism, he chose to consider the cosmological constant an irreducible constant of nature: "After all the imagining of mechanisms has been out of fashion this last quarter of a century, and the behaviour of *lambda* is not more strange or mysterious than that of the constant of gravitation *kappa*, to say nothing of the quantum-constant h, or the velocity of light c." At about the same time, Tolman reflected on the future progress of cosmology, offering a remark which in retrospect is almost prophetic: "We should be greatly helped by some acceptable interpretation and determination of the cosmological constant Λ, and this might result from a successful unified field theory, or from the proper fusion of general relativity and quantum mechanics" (Tolman, 1930).

Einstein did not introduce the cosmological constant with the sole purpose of keeping the universe in a static state. As he pointed out in 1917 and at some later occasions, he also saw it as a means to avoid an unphysical negative pressure. Neglecting matter pressure Einstein found for his closed model that $R^2 = 1/\Lambda$, but if a pressure term is included the more general relationship becomes $R^2 = 1/(\Lambda - \kappa p)$, where κ is the gravitational constant $8\pi G/c^2$. Einstein recognized early on that in a formal sense the cosmological constant was equivalent to a negative pressure given by $p = -\Lambda/\kappa$, and, by 1930, he also recognized that the constant could be understood as a measure of the energy density of the vacuum. In early 1931, now aware of the expanding universe but still not ready to embrace it in either the Friedman or the Lemaître version, Einstein sketched an expanding model with constant mass density and with a cosmological constant associated with the energy density of empty space. It followed that matter must be created continually, as in the later steady-state theory. However, Einstein quickly decided that the model was no good and therefore refrained from developing it into a publishable form.

As shown by review articles from around 1932, it was known at the time that the cosmological constant can be physically interpreted as a vacuum energy density ρ_{vac} to which corresponds a negative pressure $p_{vac} = -\rho_{vac} c^2$ (see also Section 4.3.3). Indeed, it follows from the Friedman–Lemaître equations with a pressure term that

$$\rho_{vac} = \frac{\Lambda c^2}{8\pi G} \quad \text{and} \quad p_{vac} = -\frac{\Lambda c^4}{8\pi G}.$$

The vacuum energy can also be written in terms of the critical density ρ_c characterizing the Einstein–de Sitter universe. It then becomes

$$\Omega_{vac} = \frac{\rho_{vac}}{\rho_c} = \frac{\Lambda c^2}{3H^2}.$$

The equations relating the energy and pressure of the vacuum to the cosmological constant could have been written down in 1930 or even earlier, but they were first stated explicitly in an address Lemaître gave in November 1933 to the National Academy of Sciences. The address, which was principally concerned with the formation of galaxies, was published the following year. Lemaître first noted that, with the accepted mean density of matter of approximately 10^{-30} g/cm^3 distributed homogeneously throughout the universe, the total energy would be that of an equilibrium radiation at the temperature ca. 20 K. He then offered a physical interpretation of the cosmological constant:

> "Everything happens as though the energy *in vacuo* would be different from zero. In order that absolute motion, i.e., motion relative to vacuum, may not be detected, we must associate a pressure $p = -\rho c^2$ to the density of energy ρc^2 of vacuum. This is essentially the meaning of the cosmical constant λ which corresponds to a negative density of vacuum according to $\rho_0 = \lambda c^2/4\pi G \cong 10^{-27}$ g/cm^3" (Lemaître, 1934).

Lemaître did not state his value of the cosmological constant, but from the equation it comes out as $2.8 \times 10^{-44}\,\text{cm}^{-2}$, slightly outside the interval given by Tolman. As Lemaître noted in another paper in 1934, the energy density of empty space was much greater than the average density of matter (about $10^{-30}\,\text{g/cm}^3$), and the effect produced would be a cosmic repulsion proportional to the distance.

Although the vacuum energy density given by Lemaître was large compared to the matter density, it was hugely smaller than the one offered by Nernst on the hypothesis of a cosmic space filled with zero-point energy. Contrary to Nernst, Lemaître did not connect his interpretation with the zero-point energy of space or otherwise relate it to quantum theory. His paper of 1934 is today regarded an important milestone towards the recognition of dark energy, but for a long time it attracted almost no attention and failed to inspire new work related to the strange form of vacuum energy. Nor did Lemaître himself consider the paper to be very important — he never referred to it.

The negative cosmic pressure $p_{\text{vac}} = -\rho_{\text{vac}}\, c^2$ only turned up again in the early 1950s, but then in the context of steady-state cosmology and without relating it to the cosmological constant. It is worth noting that the idea of a new form of energy represented by the cosmological constant also appeared in a paper published in 1933 written by Matvei Bronstein, a 27-year-old Russian theoretical physicist. In an attempt to explain why the universe expands rather than contracts, he suggested that the cosmological constant might not be a true constant but vary in time. Falsely accused of being a counterrevolutionary and a spy, Bronstein fell a victim to Stalin's purges. He was executed by a military firing squad in early 1938.

In an interesting correspondence with Einstein from 1947, Lemaître tried in vain to convince the father of relativity theory of the necessity of the cosmological constant. But Einstein, referring to the lambda constant as "an ugly thing" which he regretted, would have nothing to do with it. Lemaître, on the other hand, considered the constant to be a most beautiful thing, not only because it solved astronomical problems (such as the age paradox) but also because it opened up new possibilities for connecting general relativity with other areas of fundamental physics. In a book published on the occasion of Einstein's 70-year's birthday, Lemaître described the cosmological constant as a "logical convenience" and a "happy accident." Although Einstein might have introduced it for the wrong reason, "The history of science provides many instances of discoveries which have been made for reasons which are no longer considered satisfactory. It may be that the discovery of the cosmological constant is such a case" (Lemaître, 1949). With regard to the vacuum energy density, Lemaître stated it to be 1.23×10^{-27} g/cm^3 or $\Omega_{vac} = 10^{-3}$, noting that "the actual value of the [matter] density is negligible; matter is, by no means, in equilibrium and the cosmical repulsion prevails on gravitation." As Lemaître saw it, the vacuum energy rooted in the cosmological constant supported his old primeval-atom model of 1931 according to which the present universe was in a state of accelerated expansion approaching the empty de Sitter world.

4.4.4. VACUUM ENERGY

The aforementioned Russian physicist and cosmologist Yakov Zeldovich, a leading figure in the development of the modern big-bang theory, kept an open attitude with respect to a non-zero cosmological constant. The possibility that the universe might evolve in agreement with Lemaître's closed stagnation model was reconsidered as the result of observations of quasars at high redshifts, which in around 1967 inspired models with a cosmological constant of the order 5×10^{-56} cm^{-2}. In part motivated by the renewed interest in Lemaître-like models, in a brief note of 1967 Zeldovich stated what had more or less been tacitly known since the 1930s: "Corresponding to the given Λ is the concept of vacuum as a medium having a density $\rho_0 = \Lambda c^2/8\pi G = 2.5 \times 10^{-29}$ g/cm^3, an energy density $\epsilon_0 = 2 \times 10^{-8}$ erg/cm^3, and a negative pressure (tension) $p_0 = -\epsilon_0 = -2 \times 10^{-8}$ dyne/cm^3" (Zeldovich, 1967). Zeldovich cautiously added that a non-zero cosmological constant was after all just a hypothesis that needed to be justified by further astronomical observations. The unit "dyne" equals 10^{-5} Newton.

Zeldovich's note of 1967 was little more than what Lemaître had said about the subject more than thirty years earlier, but in a follow-up paper he offered a more elaborate discussion in which he related the cosmological constant to the theory of elementary particles. Now more confident that the cosmological constant was real, he stated that observationally its value was limited to the range between $+10^{-54}$ and $-10^{-54}\,cm^{-2}$. More importantly, Zeldovich argued that a vacuum with a definite energy density agreed with calculations based on quantum field theory. To obtain a finite result, he had — like Nernst much earlier — to assume a highest frequency, which he did by arbitrarily assuming it to correspond to the proton's rest mass, meaning a maximum frequency of approximately 10^{24} Hz (just a few orders of magnitude above Nernst's frequency). From this, he derived $\rho_{vac} \sim 10^{17}\,g/cm^3$, and $\Lambda \sim 10^{-10}\,cm^{-2}$, tersely noting that the estimates were totally unrealistic. This was the beginning of the so-called cosmological constant problem, namely, that the value of the cosmological constant as calculated on the basis of quantum zero-point fields is hugely larger than bounds imposed by observation (see also the next section).

Most physicists and astronomers preferred to ignore the cosmological constant, but there were a few exceptions. For example, in a study of cosmological models dating from 1975, the two American astrophysicists James Gunn and Beatrice Tinsley analyzed the consequences of a negative deceleration parameter q_0, as indicated by some observations. The accelerated expansion would require a net repulsive force, and this force could only be a positive cosmological constant. "The data may call for Λ to be dusted off and inserted in the field equations," they wrote (Gunn and Tinsley, 1975). In this regard, they referred to Zeldovich's study according to which a negative pressure would "arise quite naturally out of quantum fluctuations *in vacuo*." Gunn and Tinsley argued that the most plausible model was a closed Lemaître universe with a positive cosmological constant and a Hubble constant not lower than 80 km/s/Mpc. Their paper in *Nature* carried the provocative but also prophetic title "An Accelerating Universe."

Observations could be interpreted as support of an accelerating universe with a positive cosmological constant, and the constant itself could be interpreted theoretically as an effect of the energy density of empty space. But all this was uncertain and hypothetical. Was there any experimental support for the idea that quantum vacuum fluctuations produced a zero-point pressure, such as claimed by Zeldovich? There was, but the evidence dating from

1948 was only recognized to be cosmologically relevant many years later. Referring to the cosmological constant problem, the eminent particle physicist and theoretical cosmologist Steven Weinberg, a Nobel laureate of 1979, stated in an influential review: "Perhaps surprisingly, it was a long time before particle physicists began seriously to worry about this problem, despite the demonstration in the Casimir effect of the reality of zero-point energies" (Weinberg, 1989). Named after the Dutch physicist Hendrik Casimir, the effect mentioned by Weinberg refers to a tiny attractive force between two closely spaced metal plates placed in a vacuum. Although there is nothing at all between the plates, their very presence in a vacuum will give rise to an attraction, the attractive force varying inversely proportional with the fourth power of the distance between them.

Casimir's prediction of the effect in 1948 was purely theoretical and the predicted effect so small that it was uncertain if it could ever be tested experimentally. Casimir argued in his paper that the effect was a macroscopic manifestation of the zero-point energy in empty space, but neither he nor others related it to the vacuum energy associated with the cosmological constant. In his papers from the late 1960s, Zeldovich did not refer to the Casimir effect. For many years, the effect was considered an interesting but not very important thought experiment. At the time of Weinberg's comment, it was only verified qualitatively and it took until 1996 before a quantitative and unambiguous test was achieved. Current measurements agree with theory to an accuracy of 1%, which means that the effect is real. It is generally taken as important evidence that empty space possesses a zero-point energy due to zero-point fluctuations. This view can and has been doubted, though, as the Casimir effect can be derived by methods that do not rely on the concept of zero-point energy. Nonetheless, the consensus view is that the vacuum zero-point energy is real and that it has cosmological consequences. In a certain sense, the Casimir effect has vindicated Nernst's more than century-old idea of an ether densely packed with energy.

By the late 1980s, physicists were convinced that Aristotle's *horror vacui* dictum should be replaced by the dictum that nature abhors a completely empty vacuum. They realized that the quantum vacuum must necessarily possess an energy made up of various fields, electromagnetic and others, and also that the cosmological constant was equivalent to a vacuum energy density given by $\Lambda c^2/8\pi G$. Moreover, they were generally confident that astronomical observations could best be explained by a cosmological constant equal to zero or at least exceedingly small. Unfortunately, the small value disagreed wildly with the value calculated on the basis of the standard model of elementary particles.

4.4.5. AN ACCELERATING UNIVERSE

Readers of a popular encyclopedia of cosmology published in 1996 were told that, "Even a very small cosmological constant could have an influence on how the Universe got to be the way it is, however, and some theorists have toyed with such models. ... But these models appear ugly and unnatural, requiring very careful 'fine-tuning' of the models to make them match reality" (Gribbin, 1996, p. 115). This was indeed the consensus view at the time, but a few years later it would change dramatically with the discovery of the accelerating universe driven by a new form of repulsive energy associated with the cosmological constant. Although the Nobel Prize-rewarded discovery came as a surprise, it was not completely unanticipated. For example, in a paper from 1995, the two American astrophysicists Lawrence Krauss and Michael Turner concluded that "The Cosmological Constant is Back," as the title of their paper read. According to Krauss and Turner, the best model of the universe was one with $\Omega_\Lambda = 0.6 - 0.7$, meaning that 60–70% of the critical density was made up of the energy associated with the cosmological constant. This form of vacuum energy came to be known as "dark energy," a name possibly coined by Turner in 1998.

The discovery of an accelerating, yet critically dense universe was the outcome of work done by two international research teams studying the highly redshifted light from the rare and distant supernovae. The two teams, one the Supernova Cosmology Project (SCP) and the other the High-z Supernova Research Team (HZT), at first got discordant results concerning the matter density of the universe and therefore also concerning its geometry. In 1997, the SCP reported measurements of seven supernovae that agreed with the Einstein–de Sitter model ($\Omega_m = 1$, $\Lambda = 0$), but it turned out to be a false start. It took some further work until the two teams agreed that the supernova data revealed a flat universe in a state of accelerated expansion and with only a small density of matter. Latest by 1999, the impressive consistency of the results from the two teams demonstrated beyond any doubt that we do not live in an Einstein–de Sitter universe. It followed from the measurements that dark energy as represented by the cosmological constant played a most significant role. The conclusion of the HZT group in 1998, largely shared by the competing SCP group, was that the age of the universe was approximately 14.2 billion years, that it was dominated by dark energy ($\Omega_\Lambda = 0.76$), accelerating at a speed of $q_0 = -0.75$, and that it would continue to expand eternally.

The values reported in 1998 were soon refined by combining further studies of supernovae with data obtained from studies of the fluctuations of the cosmic microwave background. By the early years of the new millennium, the results had sharpened to an age of 13.4 ± 0.2 billion years and a composition of $(\Omega_m, \Omega_\Lambda) = (0.27, 0.73)$, where ordinary matter only made up a small part of the matter density parameter Ω_m. In 2011, the Nobel Prize in physics was awarded to Saul Perlmutter (SCP), Brian Schmidt (HZT), and Adam Riess (HZT) for "the discovery of the accelerating expansion of the universe through observations of distant supernovae."

What happened in the last two years of the twentieth century was nothing less than a revolutionary change in the world picture, almost (but only almost) comparable to Copernicus' replacement of the traditional geocentric universe with the heliocentric system 455 years earlier. The new revolution created much interest in scientific circles but, contrary to the revolution in the sixteenth century, much less attention in literary and cultural circles. There were a few exceptions, though. In a short story published in 2004, the reputed American author John Updike described the accelerating universe and the effect it might have on ordinary people such as the fictional Martin Fairweather:

> "The accelerating expansion of the universe imposed an ignominious, cruelly diluted finitude on the enclosing vastness. The eternal hypothetical structures — God, Paradise, the moral law within — now had utterly no base to stand on. All would melt away. He, no mystic, had always taken a sneaky comfort in the

idea of a universal pulse, an alternating Big Bang and Big Crunch, each time recasting matter into an unimaginably small furnace, a subatomic point of fresh beginning. Now this comfort was taken from him, and he drifted into a steady state — an estranging fever, scarcely detectable by those around him — of depression" (Updike, 2005).

Updike focused on the long-term, eschatological consequences of the accelerating universe, namely that it eternally and purposelessly moves toward a state of lifeless nothingness, but scenarios of a somewhat similar kind also followed from earlier cosmological models. Much more novel, even shockingly novel, was the drastic revision of the composition of the universe. Whereas the material part of the traditional universe was made up of ordinary matter as represented by the chemical elements in the periodic table, in the new picture it was almost completely dominated by invisible yet physically real entities, namely, the mysterious dark energy and the equally mysterious cold dark matter. Ordinary matter only counted for a few percent.

With the discovery of the accelerating universe, the cosmological constant again came into the limelight of fundamental physics, this time not as a dispensable hypothesis but as a necessity. However, the cosmological constant problem first stated by Zeldovich in the 1960s remained unsolved and has only aggravated since. In modern cosmology, the mismatch between the observed vacuum energy density and the one calculated from theory has reached proportions even more grotesque than the 46 orders of magnitude suggested by Zeldovich. Some theoretical estimates result in predicted vacuum energy densities, or values of the cosmological constant, which are as much as 122 orders of magnitude greater than the vacuum energy density inferred from observations! Physicists have come up with a number of ways to resolve the problem, but so far there is no generally accepted solution.

According to standard ΛCDM cosmology, the dark vacuum energy is given by the cosmological constant, but this is not the only possibility. A minority of physicists have explored even more exotic alternatives such as "quintessence," a name which obviously alludes to Aristotle's fifth element, the celestial ether.

The hypothetical quintessence and the name for it was introduced in 1998 as an alternative to the cosmological constant. It has a negative pressure and acts as a cosmic repulsion, but its pressure is less negative and can change with time and space. It can even be positive, corresponding to attraction. Yet another possibility, known as "phantom energy" and discussed since 2003, has a larger negative pressure than the energy given by the cosmological constant and will blow up the universe to infinity within a finite span of time. Quintessence, phantom energy, and a couple of other alternatives may be theoretically possible, but they are taken much less seriously as candidates for dark energy than the constant that Einstein introduced more than a century ago and which he originally called a "for the time being unknown universal constant" (Einstein *et al.*, 1952, p. 193).

References

A

Abbott, E. A. (1998). *Flatland: A Romance of Many Dimensions*. New York: Penguin Books.

Alfvén, H. (1966). *Worlds-Antiworlds: Antimatter in Cosmology*. San Francisco: W. H. Freeman and Company.

Anderson, C. D. (1933). "The positive electron." *Physical Review* **43**, 491–494.

Anon. (1940). "Superheavy element 94 discovered in new research." *Science News Letters* **37** (22), 387.

Aristotle (1984). *The Complete Works of Aristotle*, ed. J. Barnes. Princeton: Princeton University Press.

Armbruster, P. and G. Münzenberg (1989). "Creating superheavy elements." *Scientific American* **144** (4), 66–72.

Aston, F. (1923). *Isotopes*. London: Edward Arnold & Co.

Atkinson, R. d'E. (1931a). "Atomic synthesis and stellar energy." *Nature* **128**, 194–196.

Atkinson, R. d'E. (1931b). "Atomic synthesis and stellar energy, II." *Astrophysical Journal* **73**, 308–347.

Atkinson, R. d'E. and F. Houtermans (1929). "Zur Frage der Aufbaumöglichkeiten der Elemente in Sternen." *Zeitschrift für Physik* **54**, 656–665.

B

Bahcall, J. N. (1964). "Solar neutrinos I. Theoretical." *Physical Review Letters* **12**, 300–302.

Bahcall, J. N. and H. Bethe (1990). "A solution to the solar neutrino problem." *Physical Review Letters* **65**, 2233–2235.

Baly, E. (1921). "Inorganic chemistry." *Annual Progress Report to the Chemical Society* **17**, 27–56.

Barrow, J. D. and F. J. Tipler (1986). *The Anthropic Cosmological Principle.* Cambridge: Cambridge University Press.

Beller, M. (1999). "Jocular commemorations: The Copenhagen spirit." *Osiris* **14**, 252–273.

Bergson, H. (1965). *Creative Evolution.* London: Macmillan.

Bethe, H. (1939). "Energy production in stars." *Physical Review* **55**, 434–456.

Bethe, H. and C. Critchfield (1938). "The formation of deuterons by proton combination." *Physical Review* **54**, 248–254.

Birkeland, K. (1896). "Sur les rayons cathodiques sous l'action de forces magnétiques intenses." *Archive des Sciences Physiques et Naturelles* **1**, 497–512.

Blanqui, L. (1872). *L'Éternité par les Astres: Hypothèses Astronomique.* Paris: Libraire Germer Baillière.

Bohr, N. (1913). "On the constitution of atoms and molecules." *Philosophical Magazine* **26**, 1–25, 476–502, 857–875.

Bohr, N. (1923). "The structure of the atom." *Nature* **112**, 29–44.

Bohr, N. (1981). *Niels Bohr Collected Works,* Vol. 2, ed. U. Hoyer. Amsterdam: North-Holland.

Bohr, N. (1985). *Niels Bohr Collected Works,* Vol. 6, ed. J. Kalckar. Amsterdam: North Holland.

Bohr, N. (1986). *Niels Bohr Collected Works,* Vol. 9, ed. R. Peierls. Amsterdam: North Holland.

Bond, H. *et al.* (2013). "HD 140283: A star in the solar neighborhood that formed shortly after the big bang." *Astrophysical Journal Letters* **765**, L12.

Bondi, H. and T. Gold (1948). "The steady-state theory of the expanding universe." *Monthly Notices of the Royal Astronomical Society* **108**, 252–270.

Born, M. (1920). "Die Brücke zwischen Chemie und Physik." *Naturwissenschaften* **8**, 373–382.

Born, M. (1971). *The Born-Einstein Letters.* London: Macmillan.

Boyce, N. W. (1972). "A priori knowledge and cosmology." *Philosophy* **47**, 67–70.

Boyle, L., K. Finn, and N. Turok (2018). "The big bang, CPT, and neutrino dark matter." Arxiv:1803.08930.

Brauner, B. (1923). "Hafnium or celtium?" *Chemistry & Industry* **42**, 782–788.

Burbidge, E. M., G. R. Burbidge, W. A. Fowler, and F. Hoyle (1957). "Synthesis of elements in stars." *Reviews of Modern Physics* **29**, 547–650.

C

Campbell, N. R. (1907). *Modern Electrical Theory*. Cambridge: Cambridge University Press.

Carazza, B. and H. Kragh (1995). "From time atoms to space-time quantization: The idea of discrete time, ca 1925–1936," *Studies in the History and Philosophy of Science* **25**, 437–462.

Cauchy, A. (1868). *Sept Leçons de Physique Générale*. Paris: Gauthier-Villars.

Chamberlain, O., E. Segré, C. Wiegand, and T. Ypsilantis (1955). "Observation of antiprotons." *Physical Review* **100**, 947–950.

Chandrasekhar, S. and L. Henrich (1942). "An attempt to interpret the relative abundances of the elements and their isotopes." *Astrophysical Journal* **95**, 288–298.

Charlier, C. (1925). "On the structure of the universe." *Publications of the Astronomical Society of the Pacific* **37**, 115–135.

Clausius, R. (1867). *The Mechanical Theory of Heat*. London: T. van Voorst.

Clausius, R. (1868). "On the second fundamental theorem of the mechanical theory of heat." *Philosophical Magazine* **35**, 405–419.

Couderc, P. (1952). *The Expansion of the Universe*. London: Faber and Faber.

Crookes, W. (1886). "On the nature and origin of the so-called elements." *Report of the British Association for the Advancement of Science*, 558–576.

Crowe, M. J. (1994). *Modern Theories of the Universe: From Herschel to Hubble*. New York: Dover Publications.

D

Dahl, P. (1972). *Ludvig A. Colding and the Conservation of Energy Principle*. New York: Johnson Reprint Corporation.

Darwin, C. (1958). *The Autobiography of Charles Darwin, and Selected Letters*. New York: Dover Publications.

Darwin, G. H. (1905). "President's address." *Report, British Association for the Advancement of Science*, 3–32.

De Sitter, W. (1931a). [Untitled]. *Nature* **128**, 706–709.

De Sitter, W. (1931b). "The expanding universe." *Scientia* **49**, 1–10.

De Sitter, W. (1933). [Untitled]. *Observatory* **56**, 182–185.

Dicke, R. H. (1959). "Gravitation — An enigma." *American Scientist* **47**, 25–40.

Dingle, H. (1924). *Modern Astrophysics*. London: Collins Sons.

Dirac, P. (1929). "Quantum mechanics of many-electron systems." *Proceedings of the Royal Society A* **123**, 714–733.

Dirac, P. (1930). "The proton." *Nature* **126**, 605–606.

Dirac, P. (1931). "Quantised singularities in the electromagnetic field." *Proceedings of the Royal Society of London A* **133**, 60–72.

Dirac, P. (1939). "The relation between mathematics and physics." *Proceedings of the Royal Society, Edinburgh* **59**, 122–129.

Douglas, A. V. (1956). *The Life of Arthur Stanley Eddington*. London: Thomas Nelson and Sons.

Duane, W. and G. Wendt (1917). "A reactive modification of hydrogen produced by alpha-radiation." *Physical Review* **10**, 116–128.

Duff, M. J. (2015). "How fundamental are fundamental constants?" *Contemporary Physics* **56**, 35–47.

E

Eddington, A. S. (1920a). *Space, Time and Gravitation: An Outline of the General Relativity Theory*. Cambridge: Cambridge University Press.

Eddington, A. S. (1920b). "The internal constitution of the stars." *Nature* **106**, 14–20.

Eddington, A. S. (1923). *The Mathematical Theory of Relativity*. Cambridge: Cambridge University Press.

Eddington, A. S. (1926). *The Internal Constitution of the Stars*. Cambridge: Cambridge University Press.

Eddington, A. S. (1928). "Sub-atomic energy." *Memoirs and Proceedings of the Manchester Literary and Philosophical Society* **72**, 101–117.

Eddington, A. S. (1932). "The theory of electric charge." *Proceedings of the Royal Society A* **138**, 17–41.

Eddington, A. S. (1933). *The Expanding Universe*. Cambridge: Cambridge University Press.

Eddington, A. S. (1935). *New Pathways in Science*. Cambridge: Cambridge University Press.

Eddington, A. S. (1938). "Forty years of astronomy." In J. Needham and W. Pagel, eds., *Background to Modern Science*. Cambridge: Cambridge University Press.

Einstein, A. (1929). "Space-time." *Encyclopedia Britannica* **20**, 1070–1073.

Einstein, A. (1945). *The Meaning of Relativity*. Princeton: Princeton University Press.

Einstein, A. (1982). *Ideas and Opinions*. New York: Three Rivers Press.

Einstein, A. *et al.* (1952). *The Principle of Relativity*. New York: Dover Publications.

Eisele, C. (1957). "The Charles S. Peirce — Simon Newcomb correspondence." *Proceedings of the American Philosophical Society* **101**, 409–433.

Ellis, G., K. A. Meissner, and H. Nicolai (2018). "The physics of infinity." *Nature Physics* **14**, 770–772.

Eve, A. S. (1935). "Sir John Cunningham McLennan." *Obituary Notices of Fellows of the Royal Society* **4**, 576–583.

Eve, A. S. (1939). *Rutherford*. Cambridge: Cambridge University Press.

F

Falb, R. (1875). "Die Welten, Bildung und Untergang." *Sirius: Zeitschrift für Populäre Astronomie* **8**, 193–202.

Feynman, R. P. (1985). *QED: The Strange Theory of Light and Matter*. Princeton: Princeton University Press.

Flerov, G. N. *et al.* (1991). "History of the transfermium elements Z = 101, 102, 103." *Soviet Journal of Particles and Nuclei* **22**, 453–483.

Fowler, W. A. (1984). "Experimental and theoretical nuclear astrophysics: The quest for the origin of the elements." *Reviews of Modern Physics* **56**, 172–229.

French, A. P., ed. (1979). *Einstein: A Centenary Volume*. Cambridge, MA: Harvard University Press.

Freund, I. (1904). *The Study of Chemical Composition: An Account of Its Method and Historical Development*. Cambridge: Cambridge University Press.

Friedman, A. (1922). "Über die Krümmung des Raumes." *Zeitschrift für Physik* **10**, 377–386.

Friedman, A. (2014). *The World as Space and Time*. Montreal: Minkowski Institute Press.

Friedman, A. J. and C. H. Donley (1985). *Einstein as Myth and Muse*. Cambridge: Cambridge University Press.

G

Gamow, G. (1951). "The origin and evolution of the universe." *Scientific American* **39**, 393–407.

Gamow, G. (1952). *The Creation of the Universe*. New York: Viking Press.
Gamow, G. (1954a). "On the steady-state theory of the universe." *Astronomical Journal* **59**, 200.
Gamow, G. (1954b). "Modern cosmology." *Scientific American* **190** (3), 55–63.
Gamow, G. (1965). *Mr. Tompkins in Paperback*. Cambridge: Cambridge University Press.
Gavroglu, K. and A. Simões (2012). *Neither Physics nor Chemistry: A History of Quantum Chemistry*. Cambridge, MA: MIT Press.
Ghiorso, A. and T. Sikkeland (1967). "The search for element 102." *Physics Today* **20** (9), 25-32.
Godart, O. and M. Heller (1978). "Un travail inconnu de Georges Lemaître." *Pontifica Academia delle Scienze, Commentarii* **3**, 1–12.
Gribbin, J. (1996). *Companion to the Cosmos*. London: Phoenix Giant.
Gunn, J. E. and B. M. Tinsley (1975). "An accelerating universe." *Nature* **257**, 454–457.

Heisenberg, W. (1943). "Die 'beobachtbaren Grössen' in der Theorie der Elementarteilchenphysik." *Zeitschrift für Physik* **33**, 513–538.
Helmholtz, H. (1995). *Science and Culture: Popular and Philosophical Essays*, ed. D. Cahan. Chicago: University of Chicago Press.
Hermann, G. (2014). "Historical reminiscences: The pioneering years of superheavy element research." In Schädel, M. and Shaughnessy, D., eds., *The Chemistry of Superheavy Elements*, pp. 485–510, Berlin: Springer.
Herschel, J. (1833). *Treatise on Astronomy*. London: Green & Longman.
Herzberg, G. (1979). "A spectrum of triatomic hydrogen." *Journal of Chemical Physics* **70**, 4806–4807.
Herzberg, G. (1985). "Molecular spectroscopy: A personal history." *Annual Review of Physical Chemistry* **36**, 1–31.
Hilbert, D. (2013). *David Hilbert's Lectures on the Foundation of Arithmetic and Logic 1917–1933*, eds. W. Ewald and W. Sieg. Heidelberg: Springer.
Hirschfelder, J. (1938). "The energy of the triatomic hydrogen molecule and ion." *Journal of Chemical Physics* **6**, 795–806.
Hoyle, F. (1986). "Personal comments on the history of nuclear astrophysics." *Quarterly Journal of the Royal Astronomical Society* **27**, 445–453.
Hoyle, F. and N. C. Wickramasinghe (1999). "The universe and life: Deductions from the weak anthropic principle." *Astrophysics and Space Science* **268**, 89–102.

Hubble, E. P. (1929). "A clue to the structure of the universe." *Astronomical Society of the Pacific Leaflets* **1**, 93–96.

Hubble, E. P. (1942). "The problem of the expanding universe." *Science* **95**, 212–215.

Hufbauer, K. (1981). "Astronomers take up the stellar-energy problem." *Historical Studies in the Physical Sciences* **11**, 277–303.

J

Jeans, J. (1901). "The mechanism of radiation." *Philosophical Magazine* **2**, 421–455.

Jeans, J. (1913). "Discussion on radiation." *Report, British Association for the Advancement of Science*, 376–386.

Jeans, J. (1928a). "The physics of the universe." *Nature* **122**, 689–700.

Jeans, J. (1928b). "The wider aspects of cosmogony." *Nature* **121**, 463–470.

Jordan, P. and W. Pauli (1928). "Zur Quantenelektrodynamik ladungsfreier Felder." *Zeitschrift für Physik* **47**, 151–173.

K

Kant, I. (1981). *Universal Natural History and Theory of the Heavens*, trans. S. L. Jaki. Edinburgh: Scottish Academic Press.

Kant, I. (2004). *Metaphysical Foundations of Natural Science*, ed. M. Friedman. Cambridge: Cambridge University Press.

Kayser, H. (1910). *Handbuch der Spektroskopie*, Vol. 5. Leipzig: Hirzel.

Kragh, H. (1989a). "Concept and controversy: Jean Becquerel and the positive electron." *Centaurus* **32**, 203–240.

Kragh, H. (1989b). "The negative proton: Its earliest history." *American Journal of Physics* **57**, 1034–1039.

Kragh, H. (1990). *Dirac: A Scientific Biography*. Cambridge: Cambridge University Press.

Kragh, H. (1993). "Between physics and chemistry: Helmholtz's route to a theory of chemical thermodynamics." In D. Cahan, ed., *Hermann von Helmholtz and the Foundations of Nineteenth-Century Science*, pp. 401–429. Berkeley: University of California Press.

Kragh, H. (2008). *Entropic Creation: Religious Contexts of Thermodynamics and Cosmology*. Aldershot: Ashgate.

Kragh, H. (2010). "Auroral chemistry: The riddle of the green line." *Bulletin for the History of Chemistry* **35**, 97–104.

Kragh, H. (2012). *Niels Bohr and the Quantum Atom: The Bohr Model of Atomic Structure 1913–1925*. Oxford: Oxford University Press.

Kragh, H. (2016). *Julius Thomsen: A Life in Chemistry and Beyond*. Copenhagen: Royal Danish Academy of Sciences and Letters.

Kragh, H. (2017). "The Nobel Prize system and the astronomical sciences." *Journal for the History of Astronomy* **48**, 257–280.

Kramers, H. and H. Holst (1923). *The Atom and the Bohr Theory of its Structure: An Elementary Presentation*. London: Gyldendal.

L

Lakatos, I. (1970). "Falsification and the methodology of scientific research programmes." In I. Lakatos and A. Musgrave, eds., *Criticism and the Growth of Scientific Knowledge*, pp. 91–196. Cambridge: Cambridge University Press.

Landau, L. (1932). "On the theory of stars." *Physikalische Zeitschrift der Sowietunion* **1**, 285–288.

Landolt, H. H. (1893). "Untersuchungen über etwaige Aenderungen des Gesamtgewichtes chemisch sich umsetzender Körper." *Zeitschrift für Physikalische Chemie* **12**, 1–32.

Larmor, J. (1894). "A dynamical theory of the electric and luminiferous medium, I." *Philosophical Transactions of the Royal Society* **75**, 719–822.

Lemaître, G. (1931a). "The beginning of the world from the point of view of quantum theory." *Nature* **127**, 706.

Lemaître, G. (1931b). "L'expansion de l'espace." *Revue Questions Scientifiques* **17**, 391–410.

Lemaître, G. (1933). "L'univers en expansion." *Annales de Sociétés Scientifique de Bruxelles* **53**, 51–85.

Lemaître, G. (1934). "Evolution of the expanding universe." *Proceedings of the National Academy of Sciences* **20**, 12–17.

Lemaître, G. (1949). "The cosmological constant." In P. Schilpp, ed., *Albert Einstein: Philosopher-Scientist*, pp. 437–456. Evanston: Library of Living Philosophers.

Lenz, W. (1926). "Das Gleichgewicht von Materie und Strahlung in Einsteins geschlossener Welt." *Physikalische Zeitschrift* **27**, 642–645.

Lesser, L. M. (2006). "Hotel Infinity." *American Mathematical Monthly* **113**, 704.

Lewis, G. N. (1922). "The chemistry of the stars and the evolution of radioactive substances." *Publications of the Astronomical Society of the Pacific* **34**, 309–319.

Lindsay, R. B. (1973). *Julius Robert Mayer: Prophet of Energy*. Oxford: Pergamon Press.

Lobachevsky, N. I. (1898). *Nikolaj Iwanowitsch Lobatschefskij: Zwei geometrische Abhandlungen*, ed. F. Engel. Leipzig: Teubner.

Lodge, O. (1907). "The density of the æther." *Philosophical Magazine* **13**, 488–506.

Lodge, O. (1924). *Atoms and Rays: An Introduction to Modern Views on Atomic Structure and Radiation*. New York: George H. Doran Co.

Lodge, O. (1925). *Ether & Reality*. New York: George H. Doran Co.

Lucretius (1904). *On the Nature of Things*. Amherst, NY: Prometheus Books.

M

Mach, E. (1923). *Populärwissenschaftliche Vorlesungen*. Leipzig: J. A. Barth.

Martin, D. W., E. W. McDaniel, and L. M. Meeks (1961). "On the possible occurrence of H_3^+ in interstellar space." *Astrophysical Journal* **134**, 1012–1013.

Maxwell, J. C. (1973). *The Scientific Papers of James Clerk Maxwell*, ed. W. D. Niven. New York: Dover Publications.

McMillan, E. (1951). "The transuranium elements: Early history." https://www.nobelprize.org/uploads/2018/06/mcmillan-lecture.pdf.

Mendeleev, D. I. (1889). "The periodic law of the chemical elements." *Journal of the Chemical Society* **55**, 634–656.

Mendeleev, D. I. (1904). *An Attempt Towards a Chemical Conception of the Ether*. London: Longmans, Green & Co.

Meyer, L. (1872). *Die Modernen Theorien der Chemie und Ihre Bedeutung für die Chemische Statik*. Breslau: Maruscke & Berendt.

Millikan, R. A. (1917). *The Electron*. Chicago: University of Chicago Press.

Millikan, R. A. (1951). *The Autobiography of Robert A. Millikan*. London: MacDonald.

Milne, E. A. (1948). *Kinematic Relativity*. Oxford: Clarendon Press.

N

Nernst, W. (1916). "Über einen Versuch, von quantentheoretischen Betrachtungen zur Annahme stetiger Energieänderungen zurückzukehren." *Verhandlungen der Deutschen Physikalischen Gesellschaft* **18**, 83–116.

Newcomb, S. (1878). *Popular Astronomy*. New York: Harper and Brothers.

Newcomb, S. (1898). "The philosophy of hyper-space." *Science* **7**, 1–7.
Newton, I. (1952). *Opticks*. New York: Dover Publications.
Nicholson, J. (1913). "The physical interpretation of the spectrum of the corona." *Observatory* **36**, 103–112.

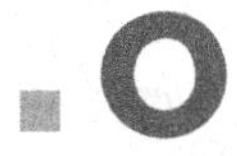

O'Brien, D. (1981). *Theories on Weight in the Ancient World*. Paris: Brill.
Oka, T. (2011). "Spectroscopy and astronomy: From the laboratory to the Galactic center." *Faraday Discussions* **150**, 9–22.
Olbers, H. W. (1826). "Ueber die Durchsichtigkeit des Weltraums." *Astronomisches Jahrbuch* **51**, 110–121.
Öpik, E. J. (1960). *The Oscillating Universe*. New York: New American Library.

P

Pais, A. (1991). *Niels Bohr's Times, in Physics, Philosophy, and Polity*. Oxford: Clarendon Press.
Paneth, F. (1916). "Über den Element- und Atombegriff in Chemie und Radiologie." *Zeitschrift für Physikalische Chemie* **91**, 171–198.
Park, B. S. (1999). "Chemical translators: Pauling, Wheland and their strategies for teaching the theory of resonance." *British Journal for the History of Science* **32**, 21–46.
Pauling, L. (1939). *The Nature of the Chemical Bond*. Ithaca, NY: Cornell University Press.
Pell, M. (1872). "On the constitution of matter." *Philosophical Magazine* **43**, 161–185.
Planck, M. (1911). *Vorlesungen über Thermodynamik*. Leipzig: Veit & Co.
Planck, M. (1960). *A Survey of Physical Theory*. New York: Dover Publications.
Pynchon, T. (1874). *Introduction to Chemical Physics*. New York: Van Nostrand.

R

Ramsay, W. (1908). "The electron as an element." *Journal of Chemical Society* **93**, 774–788.
Ramsay, W. (1909). *Essays — Biographical and Chemical*. London: Constable & Co.

Riemann, B. (1873). "On the hypotheses which lie at the bases of geometry." *Nature* **8**, 15–17, 36–37.

Russell, B. (1897). *An Essay on the Foundations of Geometry*. Cambridge: Cambridge University Press.

Russell, H. N. (1929). "The highest known velocity." *Scientific American* **140** (6), 504–505.

Rutherford, E. (1904). *Radio-Activity*. Cambridge: Cambridge University Press.

S

Sandage, A. R. (1957). "The red shift." In G. Piel, *et al.*, eds., *The Universe*, pp. 89–98. New York: Simon and Schuster.

Sandage, A. R. (1958). "Current problems in the extragalactic distance scale." *Astrophysical Journal* **127**, 513–526.

Schilpp, P., ed. (1949). *Albert Einstein, Philosopher-Scientist*. New York: Library of Living Philosophers.

Schultze, G. R. (1931). "Triatomic or monatomic hydrogen." *Journal of Physical Chemistry* **35**, 3186–3188.

Schuster, A. (1898). "Potential matter — a holiday dream." *Nature* **58**, 367.

Schwarzschild, K. (1900). "Über das zulässige Krümmungsmass des Raumes." *Vierteljahrschrift der Astronomischen Gesellschaft* **35**, 337–347.

Scriven, M. (1954). "The age of the universe." *British Journal for the Philosophy of Science* **5**, 181–190.

Segré, E. (1980). *From X-Rays to Quarks: Modern Physicists and Their Discoveries*. San Francisco: W. H. Freeman.

Shrum, G. (1986). *Gordon Shrum: An Autobiography*. Vancouver: University of British Columbia Press.

Singer, D. (1950). *Giordano Bruno: His Life and Thought*. New York: Henry Schuman.

Smithells, A. (1907). "Presidential address." *Nature* **76**, 352–357.

Smolin, L. (2004). "Atoms of space and time." *Scientific American* **290** (1), 66–75.

Soddy, F. (1904). *Radioactivity, An Elementary Treatise*. London: The Electrician.

Stark, J. (1917). "Erfahrung und Bohrsche Theorie der Wasserstoffspektren." *Annalen der Physik* **54**, 111–116.

Steinhardt, P. (2004). "The endless universe: A brief introduction." *Proceedings of the American Philosophical Society* **148**, 464–470.

T

Tait, P. G. (1876). *Lectures on Some Recent Advances in Physical Science.* London: Macmillan.

Tegmark, M. (2003). "Parallel universes." *Scientific American* **288** (5), 41–51.

Thomson, J. J. (1897). "Cathode rays." *Proceedings of the Royal Institution,* 1–14.

Thomson, J. J. (1934). "Heavy hydrogen." *Nature* **133**, 280–281.

Thomson, W. (1852). "On a universal tendency in nature to the dissipation of mechanical energy." *Philosophical Magazine* **4**, 304–306.

Thomson, W. (1891). *Popular Lectures and Addresses,* Vol. 1. London: Macmillan.

Tolman, R. C. (1922). "Review of the present status of the two forms of quantum theory." *Journal of the Optical Society of America and Review of Scientific Instruments* **6**, 211–228.

Tolman, R. C. (1930). "Discussion of various treatments which have been given to the non-static line element for the universe." *Proceedings of the National Academy of Sciences* **16**, 582–594.

Tolman, R. C. (1932). "Models of the physical universe." *Science* **75**, 367–373.

Tolman, R. C. (1934). *Relativity, Thermodynamics and Cosmology.* Oxford: Oxford University Press.

Trimble, V. (1988). "Dark matter in the universe: Where, what, and why?" *Contemporary Physics* **29**, 373–392.

U

Updike, J. (2005). "The accelerating expansion of the universe." *Physics Today* **58** (4), 39.

Urey, H. C. (1933). "Editorial." *Journal of Chemical Physics* **1**, 1–2.

Urey, H. C. (1934). "Some thermodynamic properties of hydrogen and deuterium." https://www.nobelprize.org/prizes/chemistry/1934/urey/lecture/.

V

Van Vleck, J. H. (1928). "The new quantum mechanics." *Chemical Reviews* **5**, 467–507.

Vegard, L. (1924). "The auroral spectrum and the upper atmosphere." *Nature* **113**, 716–717.

Veneziano, G. (2004). "The myth of the beginning of time." *Scientific American* **290** (5), 54–65.

W

Wapstra, A. H. *et al.* (1991). "Criteria that must be satisfied for the discovery of a new chemical element to be recognized." *Pure and Applied Chemistry* **63**, 879–886.
Watts, W. M. (1907). "The spectrum of the aurora borealis." *Monthly Weather Review* **35**, 405–421.
Wegener, A. (1912). "Die Erforschung der obersten Atmosphärenschichten." *Zeitschrift für Anorganische Chemie* **75**, 107–131.
Weinberg, S. (1989). "The cosmological constant problem." *Reviews of Modern Physics* **61**, 1–23.
Weizsäcker, C. F. (1938). "Über Elementumwandlungen im Innern der Sterne, II." *Physikalische Zeitschrift* **39**, 633–646.
Wendt, G. and R. Landauer (1920). "Triatomic hydrogen." *Journal of the American Chemical Society* **42**, 930–946.
Wilczynska, M. R. *et al.* (2020). "Four direct measurements of the fine-structure constant 13 billion years ago." *Science Advances* **6** (17), eaay9672.
Wilkinson, D. H. *et al.* (1993). "Discovery of the transfermium elements." *Pure and Applied Chemistry* **67**, 1757–1814.

Z

Zeldovich, Ya. (1967). "Cosmological constant and elementary particles." *JETP Letters* **6**, 316–317.
Zeldovich, Ya. and I. D. Novikov (1983). *The Structure and Evolution of the Universe.* Chicago: University of Chicago Press.
Zöllner, K. F. (1872). *Über die Natur der Cometen.* Leipzig: Engelmann.
Zwicky, F. (1935). "Remarks on the redshift from nebulae." *Physical Review* **48**, 802–806.

Index

A

B

C

D

E

F

G

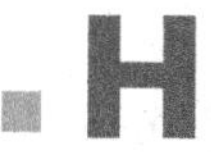

H

I

J

K

L

M

N

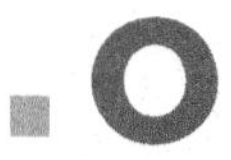

P

Q

R

S

T

U

V

W

X

Y

Z

Printed in the United States
by Baker & Taylor Publisher Services

Printed in the United States
by Baker & Taylor Publisher Services